"A wealth of information documents the probability that Noah's Ark's remains still exist on Mt. Ararat in Eastern Turkey. These include living testimony of eyewitnesses, historical accounts and archaeological discoveries. Likewise, a wealth of geologic information documents the fact of the great flood of Noah's day, and much evidence can be sited for the co-existence of dinosaurs and man. In this volume, David Larsen marries the two streams of thought into a working whole, linking them with his own personal exploration efforts. Together they present an exciting tale of investigation and discovery that will thrill all who read it."

—**Dr. John Morris**, President of the Institute for Creation Research, Dallas, TX

"In 1970, I found myself climbing Mt Lyell, the highest peak in Yosemite National Park. Little did I then realize that the peak, with its glacial snow fields strewn with granite talus as an apron around the peak, jeopardizing the unwary, was named after Charles Lyell, the "father of modern geology" who had set out to destroy the Biblical record of the flood and of Noah's Ark. Is it possible that much of what we call "science" is not, in reality, a genuine search for truth but a battle against it? What would happen if Noah's Ark were truly discovered in this momentous hour of world history? It would be the greatest archaeological discovery of all time with vast scientific and spiritual implications. You are about to embark on a fascinating journey in search of Noah's Ark. You will encounter scientific walls, philosophical and spiritual crevasses, and political intrigue while being profoundly provoked in your own thinking. Whether you are just a lover of a good read, a science buff, an adventurer, or in pursuit of truth, this book is for you. A book for our time."

—**Charles Crismier**, President, Save America Ministries, Host of *Viewpoint*, a nationally syndicated radio program

"This book is much more than a consideration of Dinosaurs being on Noah's Ark. It is a very scholarly examination of many creation/evolution issues combined with an exciting account of the author's many years of involvement in the search for the remains of the Ark on Mt. Ararat in Eastern Turkey. Evolutionary arguments are demolished with sound scientific reasoning and facts, and the excitement, mystery, intrigue and life-threatening adventures on Mt. Ararat are described in .... allow one to enter into the search, a mains indeed been found? *Dinosaurs on the Ark* is an extremely valuable resource for insight into creation/evolution issues and the on going search for the remains of the Ark of Noah on Mt. Ararat."

—**John T. McIntosh**, President-SEARCH Foundation

"If you, like me, have questions surrounding the age of the universe, the demise of the dinosaurs, the Great Pyramids, the Ice Age, and how such things fit into Scripture, and why it matters, I highly recommend this book. *Dinosaurs on the Ark* reveals David's passion for truth (and recounts the remarkable adventure of his quest to find Noah's Ark). David accurately "handles the word of truth" (2 Tim 2:15) in revealing and explaining the biblical truths of creation, and reveals the spiritual roots of the secular myth of "billions of years" of time. David's book will help the reader clearly understand how the enemy has confused the church and robbed us of our testimony, particularly on the subject of Creation. The church ought to have one clear and accurate voice in supporting Truth. If we're going to be successful in taking back the education system in this country, *Dinosaurs on the Ark* ought to be essential reading."

—**Che Ahn**, President of Harvest International Ministry

"Is the Earth a planet of great age? Did we evolve as the Darwinian model suggests? Did the dinosaurs die out some 65 million years ago? Or, are we living on a young planet? Was there a creation? Did the creation occur only a little more than six thousand years ago? Was there a global flood as the Bible teaches? What was the Ice Age? Are there a few dinosaurs still with us? The answers to questions such as these are thoroughly researched in Dave Larsen's book, and any thinking person has to honestly consider the implications to the accepted scientific thought if Dave is correct. It is my belief that he is. He has researched the records that tell us there are many proven instances world wide of dinosaur bones found jumbled together with other animals in a fashion that only a great volume of fast moving water could have caused. He brings to the forefront the ridiculous claims of evolution and of early man, and exposes the fraud. Other interesting subjects in Dave Larsen's fascinating book include mammoths...man-made constructions from a pre-flood age...and pyramid construction."

—**Richard Bright**, Author: *The Ark, a Reality?*, *Quest for Discovery*, and *Pain and Purpose in the Pacific*

DINOSAURS ON THE ARK

David Larsen

"To Mark,
It's been an honor teaching you and "mentoring" you (☺) at Oaks. Keep on pursuing Truth (Jn 14:6) & keep in touch!
God's blessings,
D Larsen

TEACH Services, Inc.
Brushton, New York

**PRINTED IN
THE UNITED STATES OF AMERICA**

World rights reserved. This book or any portion thereof may not be copied or reproduced in any form or manner whatever, except as provided by law, without the written permission of the publisher, except by a reviewer who may quote brief passages in a review. The author assumes full responsibility for the accuracy of all facts and quotations as cited in this book.

This book was written to provide accurate and authoritative information in regard to the subject matter covered. It is sold with the understanding that the publisher is not engaged in giving legal, accounting, medical or other professional advice. If legal advice or other professional expert assistance is required, the reader should seek a competent professional person.

2010 11 12 13 14 · 5 4 3 2 1

Copyright © 2009 David Larsen and TEACH Services, Inc.
ISBN-13: 978-1-57258-569-0
Library of Congress Control Number: 2008942667

Cover art by Max Stasuyk

Published by

TEACH Services, Inc.
www.TEACHServices.com

TABLE OF CONTENTS

Page:

Chapter One:
The Search for Noah's Ark Begins 1

Chapter Two:
The Spiritual Roots of Evolution Theory 19

Chapter Three:
The Scientific Age of the Earth and Universe 39

Chapter Four:
The Biblical Account of Creation 67

Chapter Five:
1998 and 1999 Searches for the Ark 92

Chapter Six:
The Biblical Account of the Flood 110

Chapter Seven:
The Geology of the Flood 131

Chapter Eight:
The Fossil Record 144

Chapter Nine:
The 2000 Search for Noah's Ark 170

Chapter Ten:
The Myth of the Ice Age 180

Chapter Eleven:
Dating a Dinosaur: What your
Mother Never Told You 194

Chapter Twelve:
The Scriptural Advent of
Animal Carnivorousness 229

Chapter Thirteen:
   The 2001 Search for Noah's Ark..........................251

Chapter Fourteen:
   Archaeological Anomalies..269

Chapter Fifteen:
   The Pre-Flood Pyramids..285

Chapter Sixteen:
   The 2003 Search for Noah's Ark..........................302

Chapter Seventeen:
   The 2005 Search for Noah's Ark..........................334

INTRODUCTION

In the second chapter of John, Jesus' first miracle is recorded as turning water into wine at a wedding feast in Cana. When the wine was taken to the steward of the wedding, he responded in essence, "Wow, this is the best wine yet! Why did you save the best for last?" Generally speaking, the best and most expensive wine would be served first, and later, after palates were dulled by a day of feasting and imbibing, the cheaper wine would be served as "filler." Often, though not always, the "best" wine is that which is carefully aged the longest. Older, in reference to wine, may not be universally best, but older is generally better, and more expensive as well. Thus, one could argue, the wine Jesus created appeared to be considerably older than it really was. So, suppose, for the sake of argument, that a modern scientist was present when the steward tasted the wine and, unfamiliar with the circumstances under which the wine had come into being, he offered to test it and help determine its age and authenticity. How old would he determine the wine to be? Ten years? Fifty years? One hundred years? Would he be able to determine the actual age, or is he dealing with completely false assumptions? The truth is, if he assumes natural formation processes, then he will be completely wrong in any attempted calculation of the wine's age or formation process. That is precisely the position that many of today's scientists, and even some biblical scholars, are in when they attempt to explain the age of the earth or universe by "natural law" or natural processes. "Science" might be able to tell us something about the formation process of turning grapes into wine (by natural process), but the problem is that science can tell us nothing about the supernatural process of turning water into wine. Neither can scientific method shed any light on the age of a universe supernaturally created from nothing.

Many times I have had well-meaning Christian scientists and/or theologians tell me that (for instance) since starlight appears to have come from stars that are apparently billions of light years away, the universe must be billions of years old. Otherwise, they say, God the Creator would have been deceptive in creating a "young" universe that has some appearance of age. Putting aside (for now) all of the poor "science" and major philosophical assumptions that have gone into the determinations of "billions of light years of distance" (which is completely unverifiable by any process of science—see, for example, the chapter on "The Scientific Age of the Earth and Universe"), my response is that, in that case, all miracles are necessarily deceptive and therefore impossible. Was not the wine that was miraculously transformed from water a "deception"? It appeared to be fine, well-aged wine when in reality it was formed from jars of water only moments before. The steward was certainly initially "deceived." Any broken bone that is instantly healed would be a theological impossibility because, given natural time constraints and laws of nature, it would appear to have taken months of gradual processes, and it would therefore be "deceptive" on God's part. But tell that to the person healed: "You've been deceived my friend!" The miracle of turning water into wine cannot be considered "deceptive" at all precisely because it is recorded as a miracle. It was done in the open for all to see (except, of course, the steward whom we assume was quickly set straight—rumors like that would have quickly spread amongst the guests). It was revealed to be a miracle—an unnatural or supernatural process. We are told how long it took. There is no deception because it is revealed to us to be a creative, supernatural event. If the wine had been formed from water in a secret back room and no one had been told how it was done, then one might have a valid argument that it was "deceptive," but the problem is that we would never have had the slightest inkling that such a thing took place. The wedding guests would indeed have been deceived, but we would never have known. The story would never have been recorded. It would simply have been another routine wedding where guests drank wine and no one would have been the wiser.

The point is, that if scientists had been at the wedding in Cana and they chose to use natural laws of physics, chemistry and biology to "date" the wine, which would we trust: the interpretation of scientists, or the Revelation of God? Just as we could not use natural laws or processes to give an accurate age of the miraculously created wine, neither can we use the prognostications of scientists to tell us the true history of the formation of the Universe or its age. No scientist was there, it is a non-repeatable event, and the Bible clearly describes it as a supernatural creative act of the Creator (not an initiation of "natural processes").

Many evolutionists and/or philosophical naturalists have tried to caricature proponents of creation science by labeling them as irrational, "flat-earth believers" (never mind that it was never a teaching of the Church—or inferred from Scripture—that the earth is flat!). Illustrations are used of primitive, ignorant, and superstitious people who (for instance) thought that thunder was the voice of the gods, or that lightning was the thunderbolt of Zeus. Now, so the illustration goes, we have "scientific" understanding of the natural processes that produce thunder and lightning, and we can reject the foolish notions of gods entirely. It is apparently irrelevant that no modern creationist has ever argued that every "natural" process is really a supernatural event or miracle. The very word "miracle" bespeaks an unusual event that supersedes or defies natural law and processes. A miracle requires that "natural laws" not only exist, but that they are the normal order of things, business as usual, a constant "modus operandi." If they were not, a "miracle" could never have been identified at all. The word would not even exist. Without natural law and processes to measure them against, we would have no bearing for recognizing the miraculous. That fact aside, perhaps even more relevant is the fact that knowing how something works "naturally" says nothing about where it came from in the first place.

For example, there are many "natural" processes that take place in an internal combustion engine: a gas-air mixture ignites from a spark, electrical impulses move along wires according to electromagnetic properties and laws, momentum carries pistons up and down, rotational inertia helps keep the crankshaft turning and valves moving, events of heating and cooling all operate according to the natural laws of thermodynamics, and so on. A mechanic can study each aspect of the engine and learn all about the natural laws in operation. He can learn what the engine does and even why. Once he has learned the way things are "supposed" to work, he can become an expert in diagnosing abnormalities when they occur. However, knowing that an electrical impulse at exactly the right moment will cause the spark plug to ignite the fuel mixture which will cause the piston to move downward making the crankshaft turn and the engine to run, tells him *nothing* about where the spark plug—or the engine—came from in the first place—or when. The entire system operates on "natural laws"—the laws of thermodynamics, electromagnetism, inertia, momentum, and so on—but to think that that necessarily means that the system was the product of those same natural laws is ludicrous. No one with eyes and a modicum of intelligence could look at a gasoline-powered engine and not know in an instant that it was very carefully designed by intelligent and highly creative engineers. Natural law enables us to determine how some things work, not where they came from in the first place. And it is the operation of those very same natural laws that convince us that the engine that operates *according to natural law* could never have been formed *by those laws*. The same is true of the universe—the natural laws that are constantly in operation tell us absolutely nothing about the initial formation of any of it.

Modern evolutionists like to make lofty-sounding arguments decrying the admission of the "miraculous"

Introduction

into nature: "If we allow for the occurrence of even one miracle," so go the arguments, "then everything could be a miracle and there is no 'natural' and science cannot function." Of course, that is tantamount to saying that because the appearance of most cliff faces can be seen to be the product of natural erosion processes over time, that all cliff faces *must* be explained by the same processes. Therefore, the faces of the presidents on Mt. Rushmore must be a result of erosion over time. Quite obviously, such an assumption would be ridiculed as nonsensical and foolish by any intelligent observer (and even most unintelligent ones!). It is precisely because we can see what natural erosion is capable of—the observation and operation of "natural law" and processes—that we can easily determine that Mt. Rushmore could never have been the product of these "natural" processes. It is not natural law or "science" that eliminates the need for the supernatural. On the contrary, it is understanding natural law that enables us to realize the absolute necessity of a Supernatural Creator for the incredible complexities of living systems and creatures (complex and intricate far beyond, by the way, the ability of our best scientific minds to even duplicate, let alone rival).

Colin Patterson, Senior Paleontologist of the British Museum of Natural History, addressed a symposium of some of the world's leading evolutionists in 1981 and asked them to answer one question: "Can you tell me anything you know about evolution," he asked, "any one thing, any one thing that is true?" Incredibly, even amongst their peers and fellow believers in evolution, not one scientist was willing to stand and declare even one thing they knew to be true about evolution, until, after a long, embarrassing silence, one man stood and said, "I do know one thing—it ought not to be taught in high school!" And yet, over twenty-five years later, evolution theory, an apparent embarrassment to many of its own proponents when challenged for evidence of its veracity, continues to be the exclusive teaching on origins in all of our public schools—and many of our private

Christian schools as well! If we were to insist in our schools that only "natural" processes (earthquakes, rain, erosion) caused the apparent faces on Mount Rushmore (infinitely less complex than a real, human head), we might come up with some unusual—and even creative!—theories. But, in the end, they would all be utter nonsense. And everyone would know it! For too long we have been "oohing and aahing" over the "clothes" of evolution (evolutionary "finds" like Piltdown Man, Java Man, Peking Man, and many other fictional and farcical "missing links"—see especially chapter eight for illustrations from the fossil record). It is high time someone shouted, "The emperor [of evolution] has no clothes!"

It is the thesis of this book, that the only proven, completely reliable history of the earth, its creation and God's place in it, is the Bible. Occasionally, archaeology, "history," geology, and even some "scientific" claims have seemed to contradict the Bible, only to be later proven false. On the contrary, the best of science, history, geology and archaeology have completely validated the biblical claims over and over. But one claim in particular has been a lightning rod for collecting the criticisms, ridicule, and even ire of geologists, archaeologists and historians alike, and that is the biblical claim that the earth and all its inhabitants were once destroyed in a global flood; the Flood of Noah. Not even the Incarnation and Resurrection have received such ridicule.

The discovery and documentation of Noah's Ark would serve to confirm the authenticity and accuracy of not only the biblical account of history in general, but more specifically, one of the most ridiculed stories in all of Scripture. The confirmation of the story of Noah would demonstrate that the world was once inundated by a global flood, and that alone would consequently wreak havoc with the current secular views of history, geology, archaeology, paleontology, and especially evolution theory. If it can be shown that there was indeed a global flood that "covered all the highest mountains on earth," most of modern geology and all of modern

Introduction

paleontology would be relegated to the realm of fiction. The false premises upon which they are currently constructed would be exposed.

Tying together the potential discovery of Noah's Ark with its ramifications for the fossil record, the biblical account of creation and Earth's true history, this book seeks to restore the full veracity of biblical truth to our churches and schools, as well as to lay bare the false assumptions and tenets of evolution theory.

Consequently, in addition to demonstrating the biblical truths of creation and the flood, as well as supporting those claims through historical, archaeological and geological sources, this book also recounts my firsthand efforts to find and document the remains of Noah's Ark. Intriguing and hard-to-overlook eyewitness accounts continue to place the Ark of Noah high on Mt. Ararat in eastern Turkey. Such a discovery could very well signal the beginning of the end of the stranglehold that evolutionary teachings have over the nation's schools. Gilbert H. Grosvenor, former editor of National Geographic magazine, once admitted that finding the remains of Noah's Ark would "be the greatest archaeological find in human history and the greatest event since the Resurrection of Christ; and [that it would] *alter all the currents of scientific thought.*" That is indeed my prayer. And the purpose of this book.

Chapter 1
THE SEARCH FOR NOAH'S ARK BEGINS

When I started college, I was a Christian and I believed in the overall truth of the biblical account of history. Yet I had been taught in the public school system that the world was billions of years old, that dinosaurs became extinct millions upon millions of years ago, that "cavemen" and other "primitive" life forms existed long before mankind appeared on the scene. Somehow, I assumed that God was still the Creator; but when and how he created was just not all that important, even if "reality" did not exactly mesh with the biblical account of history and creation. The truth is, I had never really intentionally or closely examined the biblical account of creation, so I had no idea whether it would allow for the stories I was told elsewhere. All of that was to change significantly when I got to college.

My first roommate, Dale, had begun college a year before me as a science major at a public university in Oklahoma, then, as a Christian, transferred to a private Christian university (Oral Roberts University) to go into ministry. Dale had done some reading and studying on scientific evidences for both creation and the Flood of Noah. We got into several discussions and I would try to argue that since the fossils seemed to occur in layers of different types of organisms (not that I had ever actually studied them), each layer might represent different time periods—or even unique acts of creation. Dale, instead, pointed out that water was known to have tremendous sifting capacity and the many aspects and events of a massive, yearlong, global flood could account for the different layers deposited rapidly, one on top of the other. He brought up things that had never occurred to me. For instance, instead of separate events or time periods, the so-called geological layers could represent different ecological environments or separate events within one flood. Our discussions challenged me to think, and to examine the issues at stake more thoroughly.

Initially, I saw no reason for the "days" of creation in the biblical account to be literal, twenty-four-hour periods of time and I had assumed the fossil layers represented vast eras of time. Of course, I had not studied closely either the meaning of the Hebrew text or the evidence of the fossil record, so I was largely arguing from ignorance—a fact that my roommate periodically exposed during his side of the discussion. I eventually realized that even if we allowed for the fossil layers to stem from different events, they still had to represent massive destructive events. They were evidence of death, not of life—or "creation"—and certainly not of evolution. In fact, the geologic "layers" almost always appeared in complete conformity with each other, with no hint of time-separation or erosion between them. Furthermore, fossils—especially trees—were often found extending through many layers (and even upside down). That seemed to be a clear impossibility if these layers were separated by millions of years of time. My creation paradigm—never adequately researched in the first place—was being shaken.

However, one day, Dale urged me to read a book about the search for Noah's Ark, by Montgomery. It was this book that really captivated me and would not let go. The idea that the Ark of Noah might still exist drew me like a magnet. As I read the book, I realized that I would give anything to be a part of the team that actually found and documented the ark. To be able to examine the various cages, the feeding and watering systems, and the methods of construction used by Noah himself was a dream almost beyond belief. And to be able to demonstrate the veracity of the Bible to confused and deceived high school and college students was becoming a heartfelt longing. Yet as I began to read of the many people who had tried and "failed" in their attempts to find the ark, it was a dream that I was to put on the shelf for almost twenty years. My thinking at the time was "If not them, then why me? Who am I to think that I would be any different?" Did I have a special unction from God that the

The Search for Noah's Ark Begins

others did not? I was not a theologian or an archaeologist or a geologist by training. Nevertheless, it was a recurring dream that simply would not give me rest.

And as I went on in college and later in graduate school, I began to see even more clearly how crucial the history of the Noahic Flood is—not only for establishing the absolute veracity of Scripture in every detail, but for opposing the false assumptions of evolution theory. The story of a global flood is hugely significant. As I studied and even began to look at the roots of evolution theory, I began to realize that evolution was an idea that (though apparently now embraced by many who consider themselves Christians) was at its roots diametrically opposed to the creation account of Scripture. In fact, Charles Lyell, considered by many to be the "father of modern geology," called the idea of a global flood an "incubus" to the study and interpretation of geology. He also wrote that one of his purposes in interpreting the features of nature and geology as he did (to vastly lengthen the time periods needed for formation) was to "drive men out of the Mosaic record." (See chapters "The Spiritual Roots of Evolution Theory" and "The Myth of the Ice Age.")

In other words, Lyell intentionally ruled out the possibility of a global flood before his investigation, and then interpreted everything he saw so as to turn attention away from that very possibility. That is hardly an acceptable scientific methodology, even if a global flood had not occurred, but especially if it had! If there was a global flood, Lyell's interpretation of geological features was designed to produce false conclusions and even deception. In fact, even if there had been a global flood, Lyell was determined to eradicate it from memory and the history books. Scripture prophesies of just such a scenario: "In the last days scoffers will come, scoffing and following their own evil desires ... But they deliberately forget that long ago by ... waters ... the world of that time was deluged and destroyed" (II Peter 3:3–6).

Sadly, for me, the intentional deception inherent in the roots of Lyell's teachings was made even more apparent when I heard from my former roommate after nearly twenty years of no contact. In college, he was so enamored by the search for the ark that he contacted John Morris of the Institute for Creation Research to find out how he could best become part of the search. John encouraged him to get an advanced degree in the geological sciences. After graduating from ORU, Dale went back to school to study geology at a secular university. In the process, I discovered, he had become convinced that the world was, indeed, billions of years old, there never had been a global flood, and that dinosaurs had died out millions of years ago. In order to account for the increased complexity of organisms in various layers of fossils, Dale had decided that aliens came from outer space to bring new genetic information. After speaking with him by phone and hearing firsthand the changes wrought by his secular geological education, I was forcefully reminded of Lyell's stated intent of "driving men out of the Mosaic record." Dale has since published his first book entitled *Read the Bible; It Will Scare the Hell Out of You. Why I Am No Longer a Christian.* Lyell's methods were not calculated to determine the real truth, but to turn people from it. And it is a method that predominates at all of our public educational institutes—high schools and universities.

As I began studying the Scriptures more in depth, I realized that many Christians have accepted a reinterpretation of Scripture (often without realizing it) that was, at its roots, founded upon anti-Christian, antibiblical principles. The flood of Noah is either "mythologized" or downplayed into a "local" event. I began to realize that the discovery of Noah's Ark would be the Rosetta Stone[1] of both scriptural interpretation and geology.

As I see it (in simplified form), there are three major and virtually inescapable conclusions achieved in verifying the remains of Noah's Ark high on Mt. Ararat: (1) There was a global flood as the Bible clearly states. It should be obvious that water cannot rise in a bubble. If there is a five-hundred-foot-long (or longer!) wooden ship on top of (or near enough) a seventeen-thousand-foot mountain, then the entire earth was covered. No one built such a boat at ground level and then dragged it to the top of a seventeen thousand-foot mountain for effect! Thus the biblical account, even the specific part about a global flood that has been so vilified by modern "science" and geology, is confirmed in a way that virtually no other find could achieve.

(2) A global flood not only verifies the biblical account, but it effectively eliminates all basis for vast eras of time on earth—one of the requisite and foundational tenets of evolution theory.[2] (See chapter "The Fossil Record.") Virtually all the vast layers of fossils on earth—which include the dinosaurs wherever they are found—must have been deposited by one flood less than five thousand years ago (it should be revealing that nothing comparable can be established as happening anywhere on earth today). The fossil layers are unique and they bespeak a unique and unrepeatable event. And since "dinosaurs" cannot be removed from the uppermost layers of fossils and then be said to predate those organisms found under them, dinosaurs and people—along with all other air-breathing land animals—had to have been on the ark together. (See chapters "Dating a Dinosaur" and "The Biblical Account of the Flood.")

(3) Lastly in sequence, though not necessarily in importance, the rediscovery of the ark would clearly establish (for anyone open-minded—or open-hearted!—enough to receive) that there is a Creator who communicates with mankind. No human undertook such a mammoth task of building a five-hundred-foot-long boat and supplying it with food and animals on a "hunch" that it might be needed in a few hundred years or so. In Southern California several years ago, many prognosticators were predicting any number of dire consequences for the coming of "El Nino" (only a few months in advance!), but I noted with interest that no one was advocating the construction of a ship big enough to save the denizens of Los Angeles. And with all the science and technology available to man, even predictions a few days or weeks in advance varied widely and were often inaccurate.

In any case, in college, I minored in physics and, in the process, took some Astronomy and Astrophysics courses. I began to be amazed at the number of things that were taught in the textbook that were entirely speculation based on a philosophy of vast eras of time and "naturalism." They were not at all based on proper application of what I had been taught as the "scientific method." Furthermore, I noticed that virtually all of the areas where explanations were difficult or forced, or even lacking entirely, were areas where observations and data did not fit easily with the underlying assumption that the universe was billions of years old. (See chapter "The Scientific Age of the Earth and Universe.") Most explanations became quite simple and straightforward when the age of the earth or universe was reduced to thousands of years.

I became more and more convinced that much of what we now call "science" was not, in reality, a search for the truth, but a battle against it. So, in 1976, I traveled to El Cajon, California, to the Institute for Creation Research to take a weeklong seminar on scientific evidence for creation. For me, it was a revelational experience. Eventually, I began teaching science and math to high school students (I did a stint as a graduate assistant in physics, teaching astronomy labs at the University of Arkansas, then graduated from Fuller Seminary in Pasadena with an M.A. in theology). I found that presenting scientific evidence for creation was one of my real passions, and always the story of the flood and the ark were paramount. My students were greatly affected. Some were challenged in their faith and beliefs. Some were changed. I realized, even more so, that teaching the truth of history was crucial.

In 1989, I began teaching physics and calculus at a Catholic girls' school in Southern California. One day, in my honors physics class, just in passing, I mentioned Noah's Ark. Spontaneously, the entire class burst out laughing. When I tried to ascertain what had happened to cause the outburst, the class became rather subdued as they realized that I had not been joking. One girl eventually sheepishly admitted to me that since the biblical account of Noah's

The Search for Noah's Ark Begins

Ark was just a fictional story, when I had referred to it in a serious tone as if I believed it to be true, they assumed I was joking. They admitted that they had been taught the story as "myth" throughout their education in parochial schools. When I began sharing information on Noah's Ark, what it meant, eyewitness accounts, and historical references—in short the historicity and veracity of the account—the whole class became so intrigued that when the lunch bell rang, they refused to leave. They stayed—and skipped their lunch—so that they could hear more. At that point, I decided, having recently met John McIntosh—ark researcher and veteran of many expeditions—I needed to finally do whatever I could to become personally involved in the search. I spoke with John by phone a few times and by letter occasionally, asking if there were any possibilities of joining him on any expeditions. At that time, because of political conditions in eastern Turkey, no teams were actually getting permission to explore Mt. Ararat, and the situation appeared bleak.

In the meantime, I began challenging the exclusive teaching of evolution theory at the school that publicly presented itself to be a "Christian" Catholic school. I reviewed the newly released "Science Curriculum Guidelines for Southern California Catholic Schools" and was amazed to find that they presented evolution as an established fact of science (without a hint of participation by the Creator). I wrote to the governing committee, challenging some of their interpretations, and asked for an opportunity to address the committee with alternatives. (One of the statements made by the guidelines was that since animals and humans both had DNA, that similarity proved that both had evolved by chance from a common ancestor. My response was to ask them a question: Since automobiles and airplanes share many features—such as an engine, a cockpit for the pilot, and an instrument panel—would that equally prove that an airplane evolved by random chance from a car? Or that they both evolved from a common source? Could it be that a car fell off a cliff, its doors opened by chance, the doors then stretched into wings from the air-resistance forces exerted, and the car became an airplane on the way down in order to survive? My point was that common features in car and airplane more plausibly illustrate the common hand of an intelligent creator.) Not only was I rebuffed in my efforts to speak before the committee, but I was eventually issued a silencing edict by the principal of the school (a nun who doubled as the head of the science department). I was thereafter forbidden to even answer students' questions on God or creation. When I said that I would not be willing to teach under such restrictions without adequate explanations, I was terminated. During the process, one of the things that I was told by the principal was that my speaking of the ark as an actual historical artifact was now opposed to Catholic doctrine. According to her, the first eleven chapters of Genesis are no longer accepted as historical by the Catholic Church—they are "myth." I have since found such a belief to be present even in many evangelical Protestant churches. I therefore became more determined than ever to find the remains of Noah's Ark.

In recent years, it also became clear to me that my dream of finding the ark was not just of my own doing, but something that God had placed within me from the time I was a young child. A few years ago, as I was scheduled to preach one Sunday morning at Vineyard Christian Fellowship in Monrovia, California, my parents came to accompany me. At breakfast, my mother said that as she was praying that morning, she felt the Lord remind her that when I was a child, there was a particular song on a recording that I wanted played for me every night before I went to bed. She did not remember all the lines, but she remembered that it was a song about God speaking unto Noah. As she spoke of it, the lines of the song began to come back to me after an interim of more than thirty years: "The Lord spake unto Noah and told him build an ark. 'Storms may come, but fear not, for, Noah, *I am nigh*. And through the upper window, you'll see me standing by.'" I wept as I remembered.

A short time later, as I approached the pulpit in the church, a woman known as prophetically gifted came to me and said that she felt the Lord had given her a word

The Search for Noah's Ark Begins

specifically for me. The word she had was "I'm not sure what this means, but I feel the Lord prompted me to tell you, 'David, *I am nigh.*'" It sounds so simple, and yet those words were the very words from the song of Noah that I had just been reminded of that morning for the first time in over thirty years. I do not pretend to understand all of what it means, and for that matter, it may ultimately have nothing to do with me being a part of the process of finding and documenting the ark. I do know, however, that my call to tell the story of Noah and to participate in the search is not just something I thought up, but something that the Creator instilled within me.

Over the years I maintained loose contact with John McIntosh and he periodically sent me updates or articles on events pertaining to the search. Finally, in 1997, I was scheduled to teach an eight-week series on creation at my church. For one of the classes, I planned to tell the story of the search for the ark. It occurred to me that it would be better to ask John if he would come and tell of his own experiences. Unfortunately, he was very busy with his own schedule, as well as his involvement with two different groups that were seeking permission to once again climb Mt. Ararat in eastern Turkey. However, he indicated that one of the teams, headed by Ken Long (and his ministry, Rock of Ages Documentaries), might have need of an expedition member. I immediately fired off a letter of introduction and then waited. Hearing nothing in response, I finally sent off some of my writings on creation and the ark. Almost immediately after that, I received a response of interest, met Ken Long, and then did some glacier training with his associate, Adeley Brewer, as well as others who were also potential participants. Then I began the process of getting equipped and financed (many friends and family members, as well as my church, Vineyard Christian Fellowship of Monrovia, generously contributed). Since actual permits were still pending, I had decided to wait until the last possible moment to book my flights as well as to cash the contribution checks I had received. In the event that the permits fell through at the last moment, I intended to return all the checks as well as the equipment I had purchased. Ken continued to assure the team members that things were set, that his connections had promised us permission, and that it was only a matter of time. Finally, one week into August, I received a call from Adeley that we needed to be in Turkey on August 17. I made my flight connections, cashed my checks, paid my bills, and packed. But things were not "set" after all.

The day before I was to leave, Ken phoned to tell me that our permits were going to be "delayed" by, perhaps, up to ten days. He wanted the team members to wait. However, most of my plane reservations, made at the last minute as they were, were for unchangeable, nonrefundable tickets. For me to cancel my flights the night before leaving would mean the loss of hundreds of dollars and the inability to purchase new tickets in ten days. I decided (after much prayer) that it was "now or never." I told Ken that I was leaving anyway and I would go to Dogubayazit to see what I could accomplish while I waited. If the permits were granted in ten days, I could delay my return flights if necessary. I knew that John McIntosh was currently in eastern Turkey and I reasoned that, once there, if anything was going to take place, I might have an outside chance of joining forces. One other team member, Mark Cutler, a geologist from Huntington Beach, California, decided to join me since he was largely in the same boat (no pun intended). Because of numerous plane delays and missed flights, I ended up providentially meeting up with Mark at the Istanbul Airport late one evening (instead of making our way separately to Ankara and meeting a day later as we had tentatively planned by phone). He came much better prepared than I did, as he actually had a book about Turkey that proved an invaluable aid in negotiating our way around the country. Without his assistance and his guidebook, I might still be wandering about some bus station in Turkey wondering, "Where am I and where do I need to go?" In any case, we got to spend the next morning exploring the sights and sounds of old Istanbul

The Search for Noah's Ark Begins

(something that would not have been on the itinerary had things gone according to Hoyle), before we flew out in the afternoon for Ankara. Somehow, when we arrived in Ankara, rather than staying in a hotel, we ended up on a bus headed for (we hoped!) Dogubayazit and Mt. Ararat.

Little did we know what we were in for (and exactly how it happened is still a bit of a blur—much as it was then for that matter). Somewhere around twenty-four hours later, and a few stops—who knows where?—in between, we pulled into Dogubayazit. Exhausted, we dragged our backpacks off the bus and into the first motel in sight. I figured we could spend one night and then leisurely look around the next day for better—or cheaper—accommodations. As Mark lay down, I found myself too restless for sleep, so I decided to wander around and look for John McIntosh.

The town was a good bit larger, with many more motels, than I had planned on, so it began to look like a more daunting task than I had first imagined. I did actually check out a few hotels (in which I quickly discovered that English-speaking locals were virtually nonexistent!) without any luck, when, providentially, I turned a corner and spotted John in the middle of the street (along with Dick Bright, whom I had not yet met). We had a pleasant reunion and later had supper together. We quickly discovered that not much was happening in the legal permit department, though he and Dick Bright had some other ventures planned (though at the time, they were, understandably, closemouthed about the details).

Mark, John, and I ended up doing a bit of sightseeing the following day, including visiting a dirt mound generally referred to as the Durupinar site. This location has been dramatically hailed by some as the remains of Noah's Ark. It is essentially a football-shaped dirt mound in the eastern Turkey hills about fifteen miles from Ararat. It received some publicity, especially in Christian circles, several years ago as some tried to make a case for this geological formation being the remains of Noah's Ark. The nature of the supporting evidences has been grossly overstated, and most who were once curious have now rejected the location as anything other than an interesting geological formation.

Two things immediately struck me when we first arrived at the Durupinar site: (1) According to the biblical account, it was approximately two and a half months after the ark landed on "the mountains of Ararat" before any other mountain peak was visible (Genesis 8:3–5). If the ark had landed in the vicinity of the Durupinar site, Mt. Ararat would have to have been visible long before they landed (unless one wants to propose that both Greater and Lesser Ararat arose only after the flood waters had receded). (2) If the ark had ever rested at this location, it would have been in plain sight and accessible for centuries! Yet there exists no hint of any historical reference to, or oral tradition of, such a thing. There are numerous such references to Mt. Ararat itself, but nothing comparable for this much more visible and easily accessible site. Whatever this mound was caused by, it certainly had nothing to do with the Ark of Noah[3] (interestingly, Mark Cutler found a seashell in the site).

The football-shaped dirt mound in eastern Turkey. Note the rock formation in the center of the mound, and the similar "wall" adjacent. [Author's photo, September 2008.]

The Search for Noah's Ark Begins

Mark and I spent a few days in the Dogubayazit area, mostly getting acquainted with Dick Bright and John and a few of the locals, but it became obvious very quickly that Ken Long's promised permits were not soon to be forthcoming (or anyone else's for that matter). Mark had friends in Izmir (biblical Smyrna) and he was anxious to visit western Turkey and the locations of the seven churches mentioned in Revelation.

The day he was set to leave, I got bushwhacked by some unknown physical malady. At the time, I did not feel nauseated; instead, I felt as if I had been run over by a steamroller. Whatever it was, I knew I was physically unable to travel. Mark, anxious as he was to see more of Turkey, graciously agreed to wait another day. As a result, that evening, I took my Bible down to the hotel lobby and sat there trying to read.

A curious Kurd wanted to know what I was reading and why. He was virtually the first local that I had encountered who spoke enough English for me to actually converse with him. As we talked a bit, he asked me why I was in Turkey. I naively told him that I was there to find Noah's Ark, because I wanted to show people that God existed and the Bible was true. Then he said, "But Americans have already found the ark" (he was referring to the Durupinar site). I smiled and responded that what Americans had found was dirt. The real ark, I said, was high on Mt. Ararat in the upper portion of the Ahora Gorge. He then said he did not believe in God because people killed each other over religious differences. I tried to explain that people killed each other because of sin, not because of God. Then I said that finding the ark would demonstrate that there is a God who desires to save people, because no one could have known that there was going to be a flood (long enough before it happened!) in order to build such a boat, unless God Himself had communicated that information. He got quiet for a minute and then said, "You're right, then I do believe there must be a God." He looked around, almost as if to see whether anyone was listening. What he said next stunned me. He said, "You are right, I have been there. I have pictures." He told me that he had grown up on the mountain as a shepherd boy and he had been everywhere. He also told me that there were several caves that had interesting things in them (I understood him to mean carvings and inscriptions). I had previously climbed into and explored the interesting Urartian cave (with the priestly figures carved outside) near Ishak Pasha Palace on the outskirts of Dogubayazit.

Ishak Pasa Palace undergoing a prolonged "restoration" process with Dogubayazit visible in the background. [Author's photo, September 2008.]

I asked him if he knew anything of its background. He told me that it had "many rooms and many interesting things" in it, and he had explored it as a youth. He claimed it had been accessible until about twenty years previously when the Turkish government began restoring the palace. He said that Kurds filled it in to keep it from the Turks. (If so, it would have been a daunting task!) Nevertheless, it is interesting to me that the walls of the cave are carved from solid rock and the cave is filled with dirt and stones that could not have come from anywhere within the cave.

The Search for Noah's Ark Begins

The "Urartian" cave in the cliff face. Note the kingly or priestly figures on either side of the opening. It requires a vertical climb of around twenty feet to access the opening. [Author's photo, September 2008.]

I then asked him if he would take me to the ark, but he simply said, "Impossible." The Kurds, he said, would not tolerate the ark falling into the hands of the Turks. I then asked him if Kurds and Americans worked together as friends and revealed the ark to the world, might it not aid the cause of the Kurds by bringing them world recognition and even assistance for their cause. He simply laughed at my naiveté. "If the ark were discovered, it would immediately be controlled by the Turkish government," he said. If the Kurds were given their independence, or "if politics changed," he said, he would gladly take me there, but not before. In the meantime, Americans and Kurds are not friends, he reasoned, because Americans sell guns and tanks to the Turks and the Turks use them to kill Kurds.

I asked him more questions about the condition of the ark and whether he had been inside it. I was curious to see whether he would make up details to further authenticate his account, but he did not. He had not been inside the ark;

he had only seen it and, he claimed, photographed it. I asked if he would simply show me a single picture of the ark. He said that to do that would reveal too much of its location and he could not do that. When I asked him if he could cover the background enough so that I could not see anything of the location, he hesitated. "Maybe," he said. "I will think about it."

Whether he gave it much thought or not, I do not know, but when Mark and I later met with him, he did not show us any photos. The following day, Mark and I headed for western Turkey on one of the most miserable bus trips I can ever recall making (if you have not yet had the experience of using a restroom in a Turkish bus stop, with no lights, with no toilet paper, with no *toilet* to sit on, while racked with diarrhea, I can only hope and pray you never do). The remainder of the trip would have been interesting, and could even have been pleasant, if it had not been for the nagging sense that I had, in one way, come so close to developing valuable information, but in the end accomplished nothing. It would not be the last time.

After returning to the States, I kept in contact with Richard Bright and John McIntosh by occasional e-mail and phone call. I let Dick know that if he ever needed an expedition partner, I was ready. Somewhere around Christmas I heard from both John and Dick that they were planning another venture, this time hoping to get a permit with Dr. Salih Bayraktutan of Ataturk University in Erzurum, and if I was willing, I could join them. I was, and I did (see chapter "The 1998 and 1999 Searches for the Ark").

But, in the meantime, I had an opportunity to speak at a couple of churches. At one small, home church, before I had even had an opportunity to say anything, a man approached me and said, "I've seen photos of Noah's Ark." At first, I was not quite sure what to make of him and, I must confess, only listened with one ear as I tried to organize my material. Since then, I have had an opportunity to become friends with him and we have spent several hours discussing his recollections (sometimes of many things besides the ark). His story, in simple form, is that his father was in the Air Force while doing photo analysis for the CIA. Sometime in 1974, his father called him into his office in order to show him some photos. In one of the photos of Mt. Ararat, he could see the prow of a large wooden ship jutting out of the ice. Snow and ice came down at an angle and covered a portion of it. His immediate response was "That's Noah's Ark!" "Sure is," said his father. The copies of the photos in his father's possession were then destroyed. He told me that the photos were taken from the SR–71 Blackbird and, to the best of his knowledge, were on the Russian side of the mountain (which would be basically northeast or upper Ahora Gorge). I found it very interesting when later leafing through John Morris's book, *Noah's Ark and the Ararat Adventure*, to come across a sketch purported to be from someone who had seen "spy plane photos" in the early 1970s. His sketch was virtually identical to the one my friend drew for me: same basic shape, same basic snow coverage, same angle; everything, except that it was a mirror image. When I eventually spoke to John Morris by phone, he admitted that he did not personally know the man who claimed to have seen the photo, and there was apparently, then, no possibility of comparing notes firsthand.

So it seemed once again that intriguing possibilities were short-circuited just prior to a potentially major breakthrough. At this point, my friend is not willing to have his name released, and according to his account, his father is not willing to discuss it or release any file-location information. As I understand it, files are coded in such a way that anyone releasing such information—even anonymously—leaves a trail clearly back to himself or herself. I have asked my friend to see if his father is willing to put information in a sealed envelope to be released to me only after he has passed away. As morbid as that may sound, it is, perhaps, the only way to uncover such information without potentially jeopardizing someone involved. In any case, whether his account or his appraisal of what he saw in 1974 is either accurate or honest, I have no way of determining. He has not sought to publish his account or gain anything from the telling, and whether anything further comes of this story remains to be seen.

An interesting side note to me is that, if his account is truthful and accurate, the "Ararat Anomaly" photographs recently released by the CIA can only be a diversionary smoke-screen intended to divert attention from the real location. It has always seemed suspect to me that the CIA now admits to having aerial photographs of an "anomaly" on Ararat (after having denied even that for decades), but claims they are not sufficiently clear to enable them to distinguish between a wooden, man-made boat (five hundred feet long!) and a natural rock formation. If that were the case, it is unlikely they could ever distinguish between a missile silo and a hole in the ground.

Endnotes

1 The Rosetta Stone is a large tablet of black basalt found in 1799 by Napoleon's men, which eventually enabled scholars to decipher Egyptian hieroglyphic characters for the first time. On the tablet were inscriptions telling the same tale in Greek as well as Egyptian demotic and hieroglyphic characters. With the Rosetta Stone, scholars had a cross-reference in a known language—Greek—with which to compare the hieroglyphic text.
2 As Charles Darwin noted in *Origin of Species*, if one could read Charles Lyell's book, *Principles of Geology*, and not admit that the world must be vastly old, one could put down *Origin* and read no further. Darwin's theory of evolution, as he himself demanded, depended absolutely on Lyell's principals of a vastly old earth and slow, gradual changes over eons of time (see chapter "The Fossil Record").
3 See Richard Bright's book, *The Ark, A Reality?* Guilderland, New York: Ranger Associates, Inc., 1989, for an interesting investigation of the possibility that there was once a lake covering this valley and another ship was sunk there.

Chapter 2

THE SPIRITUAL ROOTS OF EVOLUTION THEORY

First of all, it should be understood that evolution theory is a "theory" that is unique among all the disciplines that we have come to call "science." In fact, in addition to demonstrating the real roots of evolution theory, I hope to illustrate why it cannot be classified a science at all. The determination to present it exclusively to an entire generation of school children is a pernicious attempt to undermine the foundations of faith for society as a whole (whether that is consciously true of its advocates or not).

Whether you consider yourself a Christian or not does not change the simple fact that evolution theory was formulated by those whose primary determination was to counteract and contradict the teachings of Scripture. Therefore, evolution theory can only in any sense be true if the Bible is false. It is one or the other. The Bible or evolution. It cannot be the Bible and evolution, because they are diametrically opposed at their roots! The idea of evolution was not gleaned from a study of the biblical account of creation. In spite of numerous modern attempts at compromise, you cannot have both. The authors of the theory did their utmost to see to that.

However, before examining the roots of the theory, we need to recognize some elements of evolution theory. First of all, there appears to be an inverse relationship between actual supporting evidences and the fervor of supporting proclamations. In other words, the more evidence and observations accumulate in opposition to the theory, the more adamantly it is proclaimed as a fact, and the greater stranglehold it seems to gain on education.

C. S. Lewis once wrote in a letter to a creation advocate that "what inclines me to think you may be right in regarding [evolution] as the central and radical lie in the whole web of falsehood that now governs our lives is not so much your arguments against it as the fanatical and twisted attitudes of its defenders."[1] High school and college biology textbooks routinely present macroevolution as an established fact of science without even acknowledging its enormous shortcomings and failures. The very fact that some recent scholarly books such as Denton's *Evolution: A Theory In Crisis* and Behe's *Darwin's Black Box* have seriously attacked the very foundations of evolution theory shows that not only is the textbook position grossly overstated, it is manipulative and downright dishonest. There are enormous differences among scientists as to not only how evolution occurred (which they virtually all admit to), but whether it occurs at all (which many will not admit to). In fact, many textbooks no longer refer to the "theory" of evolution–some even refer to evolution as a "fact." If indeed evolution were an incontrovertible fact, no such statement would need to be made. No such resounding pronouncement is made about the fact that gravity pulls things to the earth, or that the moon is not made of green cheese. The very fact that such a statement about the theory of evolution is made at all demonstrates its dishonest roots.

In reality, in obscure scientific journals, in a few devastating recent books, and in private enclaves, the theory is not only questioned, but even attacked at virtually every point. And yet the public pronouncement is continually made that evolution "is a fact," and our schoolchildren are continually indoctrinated. Let me give you an example of what goes on behind closed doors. At a closed-to-the-public meeting at the American Museum of Natural History in New York in 1981, Colin Patterson, senior paleontologist at the British Museum of Natural History, made these statements:

"Last year I had a sudden realization [that] for over twenty years I had thought I was working on evolution in some way. One morning I woke up and something had happened in the night, and it struck me that I had been working on this stuff for twenty years and there was not one thing that I knew about it. That's quite a shock to learn that one can be misled for so long. [That, by the way, is one of the clues that this involves spiritual deception and not just poor science.] Either there was something wrong with me, or there was something wrong with evolutionary theory. Naturally, I know there is nothing wrong with me, so for the last few weeks I've tried putting a simple question to various people and groups of people.

"[The] Question is: Can you tell me anything you know about evolution, any one thing, any one thing that is true? I tried that question on the geology staff at the Field Museum of Natural History and the only answer I got was silence. I tried it on the members of the Evolutionary Morphology Seminar in the University of Chicago, a very prestigious body of evolutionists, and all I got there was silence for a long time [until] eventually one person said, 'I do know one thing—it ought not to be taught in high school.'"[2]

In reality, not only does evolution theory not have the support of experimental data, but the very possibility seems to be flatly contradicted by every known observation and experiment. Darwin admitted in 1860 that the fossil record, as it existed then, was the best and most obvious evidence *against* his theory. (See chapter "The Fossil Record.") Almost 150 years' worth of searching the fossil record for the myriads of transitional links predicted by Darwin has demonstrated conclusively that they really are missing (Why does it never seem to be noticed that they are also—and even more obviously—missing from the living record in the very same places?). Of the billions of required transitional forms, exactly none have been positively identified (and "positive identification" based on a fossil would never be possible in any case). And yet, today, most laymen would point to the fossil record as the primary evidence for evolution. More than anything else, this fact alone reveals the root of evolution theory as a deception that has become all-pervasive in our society.

Furthermore, experiment after experiment has verified beyond any question that life does not spontaneously generate: life comes only from life. And yet spontaneous generation remains the primary and absolutely essential tenet of evolution theory; if life cannot happen on its own, then any question of how it might change over time is irrelevant. Even with the best equipment, the participation of the most intelligent and highly trained scientists, and all the necessary ingredients, man is incapable of producing even the most "simple" of single-celled living organisms (which, by the way, are not the least bit "simple").

Nevertheless, evolutionists insist it happened without scientists or intelligent direction. Somehow it has become more believable and more acceptable as the hypothetical process is pushed further and further back into the unobservable, unrepeatable, unverifiable past. Given enough time, it is said, the impossible becomes the probable. Time is indeed the hero of the plot. (Strange, but in my own life, I have never found that the more mistakes I made over long periods of time, the better the endproduct.) And remember that it makes no difference whether you are discussing atheistic evolution or theistic evolution (as a process established by a Supreme Being), because the same visible evidence should be available for each. If God designed nature to naturally improve itself over time, that aspect of nature would be clearly discernible. Instead, we can easily observe and verify the opposite. In addition, theistic evolution is somewhat of a paradox since it attempts to insert God into a scheme that was originally designed by those who were attempting to eliminate God from nature entirely. Evolution theory was invented precisely as a naturalistic way to explain things by removing God entirely from the picture.

Mendel established laws of genetic variability and heredity that not only determined how traits are inherited, but which, if properly understood, precluded the possibility of evolution. A hundred years' worth of experiments with fruit flies have not only verified Mendel's observations, but have conclusively demonstrated the inviolability and immutability of species: fruit flies are still fruit flies and always have been. Fruit flies have been subjected to extensive research for well over a hundred years because they have a short life span and reproduce rapidly—they have been subjected to X-rays, heat, cold, radiation, and numerous other forms of torture in order to increase genetic variation and mutation. Yet, of the millions of mutated and deformed fruit flies produced over the decades, no increases in genetic information or improvements to the species can be documented.

Evolutionist Ernst Mayr once conducted an experiment with fruit flies in order to demonstrate evolution by producing either a hair-covered fruit fly, or, failing that, a hairless variety. The "normal" fruit fly has an average of thirty-six bristles, so Mayr artificially selected out the flies that contained more than that (note: this would already violate the theory of evolution because it is introducing "intelligent" direction into the process). After twenty generations he found there was an upper limit of fifty-six bristles. Any attempts to produce change beyond that parameter killed off the fruit flies. Going in the other direction, after thirty generations, he reached a lower limit of twenty-five bristles. Then, leaving the populations alone, after five years, the bristle count had returned to normal.[3] His experiment not only demonstrated the opposite of his intent, but that there are built-in parameters in the genetic information of a species that preclude the very possibility of evolution from one kind to another. As Hitching, evolutionist though he is, admits, "On the face of it, then, the prime function of the genetic system would seem to be to resist change: to perpetuate the species in a minimally adapted form in response to altered conditions, and if at all possible to get things back to normal. The role of natural selection is usually a negative one: to destroy the few mutant individuals that threaten the stability of the species."[4]

Evolutionists have often further obfuscated the issue by using the term "evolution" interchangeably with "change." This is more than just misleading, it is intentionally deceptive. Julian Huxley more honestly defines evolution as "a directional and essentially irreversible process occurring in time, which in its course gives rise to an increase of variety and an increasingly high level of organization in its products."[5] And, in fact, if human beings are the product of a system that began with a lifeless rock (earth), then there is no alternative but to accept this definition. It should be clear to even the most casual observer that this type of required change is never seen in nature apart from an intelligent designer.

Confusion is also produced by using the term "evolution" in place of both "microevolution" and "macroevolution." Microevolution is, in reality, not "evolution" at all. The term refers to very small changes that occur within a species, such as the process that produces many varieties of dogs. This is a known, testable, observable process, but it is actually the opposite process of what is required for macroevolution, or the change from one type of species into another (such as a dog changing into a horse, or, tougher yet, a fish changing into an amphibian). Virtually all "microevolution" results from a loss of genetic variability within a small portion of a specie's population. Therefore, to say that, given enough time, microevolution can produce macroevolution is not just unmerited speculation, it is demonstrably false. Microevolution is, in reality, a misnomer; it should be termed "deevolution," because it represents an overall deterioration in genetic information and variability. Millions of years of microevolution would result in macro deevolution and the complete loss of most, if not all, species altogether. Microevolution is observed and testable; macroevolution is not only unobservable, it is contradicted by every known observation and tenet of physics, biology, chemistry, and genetics. Millions of years of mistakes and loss of information do not produce an increase in either order or complexity. In fact, the continual loss of species through extinctions would seem to be a natural outgrowth of the second law of thermodynamics.

The second law of thermodynamics is one of the most fundamental principles of physics. It is so well established that it is, indeed, a *law*. It is like gravity. It has never been observed to be thwarted by natural processes. It can be clearly demonstrated by numerous experiments in a laboratory (totally unlike the theory of evolution), and it clearly contradicts the very possibility of the theory of evolution. It basically states that in any natural process, overall order (or complexity) and usable energy decrease. For instance (to use an analogy with which most are familiar), earthquakes are notorious for taking ordered, complex buildings (designed and produced by intelligent beings) and, from them, producing disordered, chaotic piles of rubble. No earthquake—or any combination of numerous earthquakes—has ever been observed to take a pile of rubble and turn it into an ordered, complex building. And yet, with evolution theory, we are told to believe, in essence, that given enough time and enough earthquakes (and storms, etc.), natural processes can produce a city. The "simplest," living, single-celled organism is now known to be complex and ordered almost beyond description—certainly exceeding in complexity any building designed and built by man. Mathematically it is far more likely that earthquakes could produce complicated buildings, with all of their wiring and plumbing included, than for natural processes to produce even one single-celled organism (ignoring the insurmountable problems inherent in making it come to life!).

But even if the unobservable, untestable process of macro-evolution were possible, given enough time, it should be abundantly clear that it could never account for the variety of organisms we find on earth today. If only because it could never work rapidly enough to overcome the processes of decay and loss that we can observe going on around us. It is tantamount to postulating that each earthquake over a magnitude of 6.0 produces one ordered building from a junkyard (without a shred of visible evidence that such a thing is even possible). It makes no difference, because what is clear is that each earthquake of the same magnitude destroys dozens, if not hundreds, of buildings. Consequently, natural processes alone will always destroy cities and never build them. We do not know of a single new species or organism that has arrived by evolutionary processes since the time of Darwin; but since that time there have been hundreds (perhaps thousands) of observed extinctions. Therefore, even if we postulate the evolution of one new organism for every hundred that are lost, it should be clear that natural processes cannot begin to account for the variety and complexity of the plant or animal kingdoms. What we observe in living organisms and systems is entirely consistent with the second law of thermodynamics and diametrically opposed to the theory of evolution.

Imagine yourself sitting at a typewriter keyboard, blindfolded, with your hands tied behind your back. Furthermore, the keys are of an unknown alphabet. For an hour or two you sit banging your head against the keyboard. I think most people would concede that, no matter how long you were able to keep banging your head, it is very unlikely that you would produce any complex patterns, let alone any intelligible literary works of genius. I think, even if you did not recognize the alphabet, that you would readily be able to discern the difference between pages typed by such a random process and those typed by an intelligent person telling a story.

Astronomers know this instinctively: those who look into outer space for signs of intelligent life look for any indication of a pattern in the radio waves being received, because they know that natural processes cannot produce order. No one could pick up a book and believe that the information contained in it was the result of random, chaotic processes. And yet many believe that the mind that created the order is, itself, the result of millions of years of random, mindless processes. If the human mind is the result of millions of years of random, chaotic, directionless processes, then how is it possible to believe that that process could be understood by the mind it produced? C. S. Lewis wondered the same thing in 1944: "The whole [evolutionary cosmology] professes to depend on inferences from observed facts. [But] unless inference is valid, the whole picture disappears ... Those who ask me to believe this world picture also ask me to believe that Reason is simply the unforeseen and unintended by-product of mindless

The Spiritual Roots of Evolution Theory

matter at one stage of its endless and aimless becoming. Here is a flat contradiction. They ask me at one moment to accept a conclusion and to discredit the only testimony on which that conclusion can be based... Granted that Reason is prior to matter and that the light of that primal Reason illuminates finite minds, I can understand how men should come, by observation and inference, to know a lot about the universe they live in. If, on the other hand, I swallow the [evolutionary] cosmology as a whole, then not only can I not fit in Christianity, but I cannot even fit in science. If minds are wholly dependent on brains, and brains on bio-chemistry, and bio-chemistry ... on the meaningless flux of the atoms, I cannot understand how the thought of those minds should have any more significance than the sound of the wind in the trees... Christian theology can fit in science, art, morality, and the sub-Christian religions. The [evolutionary] point of view cannot fit in any of these things, not even science itself. I believe in Christianity as I believe that the Sun has risen, not only because I see it but because by it I see everything else."[6]

Anyone looking at a book knows without question that the information in it was intelligently produced (or at least, designed, if not so intelligently produced ...). Yet the information present in the DNA of a single living cell exceeds the information content of entire sets of encyclopedias many times over. In fact, the DNA in each of your body's cells contains highly organized and incredibly complex information. Modern scientists are only beginning to scratch the surface of understanding its complexity. As Brown indicates, "DNA contains the unique information that determines what you look like, much of your personality, and how every cell in your body is to function throughout your life [and even one mistake could potentially be lethal]... If all the DNA in your body were [uncoiled, connected, and] placed end-to-end, it would stretch from here to the moon over 500,000 times! If all this very densely coded information were placed in typewritten form, it would completely fill the Grand Canyon 50 times! Understanding DNA is just one small reason for believing that you are 'fearfully and wonderfully made' (Ps 139:14)."[7]

Once again, not only is the evidence for evolution nonexistent, the evidence against it occurring as a natural process is overwhelming. And since evolution is obviously not taking place at a measurable rate in the observable present, millions and even billions of years are said to be required for it to occur. Yet there is, simply put, no compelling scientific reason that the earth—or universe—must be billions of years old. Indeed, there are numerous physical indicators that it is not. (See chapter "The Scientific Age of the Earth and Universe.") It is, however, an outright impossibility to read the scriptural account and—apart from "scientific" considerations—arrive at an age of 4.6 billion years for the earth. No one has even dared to pretend such a thing is possible. And if, tomorrow, scientists decide the earth must be 6.4 billion years old, that could equally well be forced into the Scriptures, but it could never be gleaned from, or predicted by, Scriptures. The biblical record, wherever it has been testable by any outside source—either archaeological or other historical sources—has never been found wanting. Most historians—secular or Christian—now concede that where the Bible can be compared with other historical accounts, it has always proven to be the more accurate. It has been stated categorically by many that no archaeological/historical find has ever contradicted or disproved the biblical account. In fact, many claims of the Bible that were once said to be in error have later been compellingly verified.

All that said, evolution theory cannot be benign! It is either the truth, whether you believe God had anything to do with it or not, or it is a lie intended to deceive even the elect if possible. Secular society has worked very hard to eradicate any mention of creation—especially a "young-earth" creation—from all school systems. So, if you believe in God and you believe evolution is the truth of the way he created things, one must wonder why Satan is not the least bit afraid of filling the minds of young people with truth. Jesus said that when you know the truth, the truth will set you free. Should not freedom come through the teaching of evolution? Jesus also said that a house divided against itself cannot stand (Mark 3:25). If God used a system to create the universe that becomes ammunition in the hands of his detractors to debunk the very creation he achieved, his efforts were not only for naught, they were counterproductive.

So how is it possible that a theory for which, even its primary authors admitted, there was no evidence (and the evidence that did exist was opposed to the theory) become so all-pervasive in society? I contend that is only possible if the roots of evolution theory are spiritual rather than rational or "scientific." Jesus said that Satan was a liar and the father of lies. Other venues have examined the history of evolution theory in depth, so total coverage is not required. It does not take much to topple a table—you only have to remove two legs. So there are two "legs" supporting the table of evolution that I want to look at briefly: Charles Lyell and Charles Darwin.

Lyell, a lawyer by training, considered himself a geologist and is often considered to be the "father of uniformitarianism." Uniformitarianism is basically a philosophical position that requires that every feature of the earth be interpreted and understood only in the light of rates and processes currently being observed. It was designed to accomplish two goals: (1) extend the age of the earth beyond the biblical parameters and (2) eliminate a global flood as a possibility. In his writings, Lyell stated that he hoped he could "drive men out of the Mosaic record." As he wrote in a letter: "Conbeare [geologist and Bishop of Bristol] admits three deluges before the Noachian! and Buckland adds God knows how many catastrophes besides, so we have driven them out of the Mosaic record fairly."[8] Consequently, his stated purpose in reinterpreting geological features was to turn men away from Scripture. The roots of Lyell's geology were thus antibiblical, and he recognized (as many modern Christians have not) that he was in a war. As he put it, "I am grappling not with the ordinary arm of the flesh, but with principalities and powers, ... for my rules of philosophizing, ... and I must put on my whole armour."[9] Lyell recognized that his was a philosophical war and not a scientific one. It was not even a matter of evidence, but of the rules that would be henceforth used for interpreting that evidence.

Lyell further wrote that the Genesis record of the flood had been an "incubus to the science of geology," and according to Hitching, "within a decade of his book's publication [1830], the date of Noah's Flood had vanished from the arena of scientific debate."[10] The clear antithesis to Lyell's position must be the understanding that if there was indeed a global flood, then all current geology must begin with that knowledge and understanding, or all of its conclusions will be grossly skewed. However, Lyell stated as his basic premise, "All theories are rejected which involve the assumption of sudden and violent catastrophes and revolutions of the whole earth, and its inhabitants ..."[11] This is not a principle of science, it is a philosophy that has at its roots the desire to overthrow the biblical account as history. Therefore, it can only, in any sense, be true if the biblical message is false.

Charles Darwin, who wrote *The Origin of Species*, had a similar philosophical perspective, and, in fact, closely followed and adhered to the writings of Lyell. He admitted that he wrote as an atheist because he saw no alternative: "I had no intention to write atheistically, but I cannot see as plainly as others do, and as I wish to do, evidence of design and beneficence on all sides of us. I cannot persuade myself that a beneficent and omnipotent God would have designedly created the ichneumonidae [a parasite] with the express intention of their feeding within the living bodies of caterpillars, or that a cat should play with mice. Not believing this, I see no necessity in the belief that the eye was expressly designed."[12]

Two things should be immediately clear: First, Darwin's "science" depends on his belief system that rejects the notion of a designer, and second, Darwin is looking at a system of nature that has been cursed by death and decay as a result of man's sin. It is not the perfect world that God either created or intended. Elsewhere, Darwin wrote that by the time he came to propose the idea of evolution—naturalistic, gradual change over immense periods of time—he had already come to the conclusion that "the Old Testament, from its manifestly false history of the world ... was no more to be trusted ... than the beliefs of any barbarian."[13] He stated that the process of his disbelief in Christianity had been slow and gradual, but it had become complete. Therefore, he wrote intrinsically, not as a scientist, but as one who was diametrically opposed to the God of creation and the message of the Bible. In fact, he called Christianity a damnable doctrine.[14] How is this a message that can now be embraced by Christians and incorporated within the Bible and Christianity?

Sadly, there is much confusion within the Church as to the meaning and content of evolution theory. I read recently that a poll taken in churches indicated that perhaps 90 percent of Christians have come to accept evolution theory in some form. My own straw polls indicate a smaller percentage than this, but I find a very large number who have incorporated some aspect of evolutionary ideas into their theology and/or belief systems without even knowing it (and the percentages among young people are even greater). The biblical message of creation is not unclear. It is, in fact, quite straightforward. The very fact that Christians feel compelled to reinterpret Scriptures based on the "science" of those who were diametrically opposed to the Truth of Scripture should make clear where the roots of evolution theory lie.

The Darwinian theory of evolution represents much more than a "scientific" explanation of changes in nature; it represents a changing worldview that begins with a rejection of the supernatural. Darwin, by the way, was far from the first to propose some concept of evolution. Many pagan philosophies begin with the same concept, and it is always in clear contrast to the story of creation asserted in the biblical text. In fact, it is at least interesting that Darwin's idea of "natural selection" was described fully, many years earlier by an ardent creationist named Blyth. Blyth, however, described natural selection as a process God established in a fallen world to preserve species in their essence, but not for establishing new species or new genetic information. Many historians believe that Darwin used this material without acknowledging its source because he did not want its origin to be associated with creation. Whatever the case in that regard, it should be clear that Darwin did not describe a new idea or concept, but he did attempt to describe a mechanism whereby it could take place. And though Darwinism has gained prominence and even been declared as "fact," it should be at least newsworthy that every single proposed mechanism—that supposedly gave the theory its unique status in the first place—has now been completely discredited by the very scientists who believed it.

It should be further noted that evolution theory did not begin as an attack on the idea of some nebulous, "Higher Power" or "Force" behind the universe. It was, at its very roots, a specific attack against the God and Creator described in the Bible. The God of Scripture is evidenced as Creator, Sustainer, Judge, and Savior. All mankind is to be ultimately held accountable to Him, and his agenda conflicts with that of self-centered, sinful humans whose tendency is—to cite Darwin—simply to "follow those impulses and instincts which are strongest." Darwin admitted not only his rejection of a Creator, but more specifically, the God of the Bible: "Disbelief crept over me at a very slow rate, but was at last complete... I can hardly see how anyone could wish Christianity was true, for if so ... my father, brother, and almost all my best friends will be everlastingly punished. And this is a damnable doctrine."[15] As biologist and medical doctor Michael Denton asserts, "The suggestion that life and man are the result of [evolution] is incompatible with the biblical assertion of their being the direct result of intelligent creative activity. Despite the attempt by liberal theology to disguise the point, the fact is that no biblically derived religion can really be compromised with the fundamental assertion of Darwinian theory. Chance and design are antithetical concepts ..."[16]

In "Confessions of a Professed Atheist," Aldous Huxley admitted that his motivation for accepting evolution theory was not exactly the scientific evidence supporting the theory: "I had motives for not wanting the world to have meaning; consequently, assumed it had none, and was able without any difficulty to find reasons for this assumption... For myself, as no doubt for most of my contemporaries, the philosophy of meaninglessness [evolution] was essentially an instrument of liberation ... from a certain system of morality. We objected to morality because it interfered with our sexual freedom."[17]

Of course, if there is no God, then Darwinism, or something like it, must be correct, but it does not take very much genius, nor do we have to look very far, to see the results of evolutionary indoctrination in our schools and on society. That is not to say that all sin has its roots in evolution theory. It is to say, however, that the theory of evolution has become a major philosophical groundwork for eliminating the word "sin" from our vocabulary. It can be, and has been, used as the basis for euthanasia, abortion, racism, homosexuality, and/or promiscuity in any form. To quote from the bible of the evolutionists, *The Humanist Manifesto:* "We believe, ... that traditional ... religions that place revelation, God, ... or creed above human needs and experience do a disservice to the human species ... we can discover no divine purpose or providence for the human species... No deity will save us; we must save ourselves... Promises of immortal salvation or fear of eternal damnation are both illusory and harmful... Modern science discredits such historical concepts ... Rather, science affirms that the human species is an emergence from natural evolutionary forces... Traditional religions are ... obstacles to human progress... In the area of sexuality, we believe that intolerant attitudes, often cultivated by orthodox religions ... unduly repress sexual conduct... Individuals should be permitted to express their sexual proclivities and pursue their life-styles as they desire."[18]

In short, a theory that has its roots in a denial of the historical, personal God of creation and the Bible, and leads to a lifestyle and society without concrete, ethical guidelines, cannot be seen as a benign theory that can either be ignored by, or incorporated within, the walls of Christianity without effect. It stems directly from the lie of the serpent. Christian author and teacher Rick Joyner says it is one of the main lies confronting the Church today—and many Christians apparently do not even know that it is to be confronted, let alone have the necessary weapons at their disposal. G. Richard Bozarth puts it this way in his essay, "The Meaning of Evolution": "Christianity has fought, still fights, and will fight science [sic] to the desperate end over evolution, because evolution destroys utterly and finally the very reason Jesus' life was supposedly made necessary. Destroy Adam and Eve and the original sin, and in the rubble you will find the sorry remains of the son of god [sic]. Take away the meaning of his death. Jesus was not the redeemer who died for our sins, and this is what evolution means, then Christianity is nothing."[19]

C. S. Lewis once said there are really only two religions in the world: Christianity and Hinduism. Hinduism, with every offshoot, is some variation of the very first lie told in the Bible—the serpent's message to Eve in the garden. It teaches that you can become a god—or like God—by your own efforts. Hinduism, Islam, Jehovah's Witnesses, Mormons, all teach some twist on the basic theme of achieving godhood on personal merit or accomplishment. The lie of evolution is basically the same lie—"We can save ourselves. We can direct our own path and our own future." The truth is, we cannot. Christianity teaches the opposite—that human beings are absolutely incapable of salvation apart from the work of God himself. The work of salvation has already been done for us on the cross through the person of Jesus Christ. Evolution theory and Christianity are not just incompatible, they are diametrically opposed at their very roots. As Denton states, "It was because Darwinian theory broke man's link with God and set him adrift in a cosmos without purpose or end that its impact was so fundamental. No other revolution in modern times ... so profoundly affected the way men viewed themselves and their place in the universe."[20] It is high time that the Church rose and with one voice proclaimed, "The emperor (of evolution) has no clothes!"

II Corinthians 10:3-6 states, "For though we live in the world, we do not wage war as the world does. The weapons we fight with are not weapons of the world. On the contrary, they have divine power to demolish strongholds. We demolish arguments and every pretension that sets itself up against the knowledge of God, and we take captive every thought to make it obedient to Christ." The battlefield is for the minds of men.

Romans 1:20 states that "since the creation of the world God's invisible qualities—his eternal power and divine nature—have been clearly seen, being understood from what has been made, so that men are without excuse." God's attributes and works are visible in nature, but only to those who accept Him. It goes on to say in Romans that with those who rejected God, "their thinking became futile and their foolish hearts were darkened. Although they claimed to be wise, they became fools and exchanged the glory of the immortal God for images made to look like mortal man and birds and animals and reptiles" (1:21-22).

Evolution theory is more than just foolishness, it is the intentional work of the enemy to confuse and deceive, "even the elect if it were possible." Creation is not only the beginning place for Scripture, it is one of the foundational tenets of the whole scriptural account. In Colossians when Paul is describing and defining the nature and character of Jesus Christ, he is constrained to begin with creation: "For by [Jesus] all things were created: things in heaven and on earth, visible and invisible... all things were created by him and for him" (Col. 1: 16). And John tells us, "In the beginning was the Word, and the Word was with God, and the Word was God... Through him all things were made; without him nothing was made that has been made" (John 1:1-3).

Some have said that it is not important whether creation took place billions of years ago or thousands; or whether it involved an evolutionary process or not. That is tantamount to saying truth is irrelevant. But Jesus said, when you know the truth, "the truth will set you free"(John 8:32). If it were unimportant, you must be willing to ask yourself why the enemy (Satan) has worked so hard to bring confusion and deceit in this area, not only to the world, but to the Church as well. Over the years, I have seen much spiritual freedom come—to Christians and non-Christians alike—through the message of creation (especially the message of a biblical, "young-earth" creation). On the other hand, I have seen much confusion and even a decreasing measure of faith produced by a compromise of Scripture with various aspects of evolution theory—and time. Many parents who do not have answers for their children in the schools choose to remain silent, bringing even more confusion to succeeding generations. In communist Russia, when the communists took the reins of power, they began by appeasing the Church: "Communism and Christianity can coexist," they said. And the Church heaved a collective sigh of relief. But the communists quickly and quietly removed the teaching of creation and all mention of God from their schools. And in their place, they taught exclusively evolution. After not too many years they went one step further and told churchgoers, "We have nothing against you or the Church per se, and as a matter of fact, you can continue going to church, but you can no longer continue taking your children with you." Separation of church and state they called it. Sound familiar? There is a reason that the enemy has worked so hard to influence children in our school systems. In each succeeding generation the lie becomes implanted a little more thoroughly until there are no longer any recognizable alternatives and the deception is complete. I have witnessed firsthand the effects on a church and even an entire denomination when a small lie or compromise is accepted. Eventually it becomes impossible to reject further compromise and further deception. And finally, the truth itself cannot be tolerated. Maybe even the Truth Himself.

Endnotes

1. Lewis, C. S. Private letter to B. Acworth reported in *The Creationists*. New York: Adolph Knopf Co., 1992, cited by Morris, Henry, ed. *That Their Words May Be Used Against Them*. Green Forest, Arkansas: Master Books, Inc., 1997, 381.
2. Patterson, Dr. Colin. (Senior Paleontologist, British Museum of Natural History, London) Keynote address at the American Museum of Natural History, New York City, 11/5/1981, cited in *The Revised Quotebook*. Australia: Answers in Genesis, 1990.
3. Hitching, Francis. *The Neck of the Giraffe: Where Darwin Went Wrong*. New Haven: Ticknor and Fields, 1982, 57.
4. Ibid., 59.
5. Huxley, Julian. "Evolution and Genetics," chap. 8 in *What is Science?* ed. J. R. Newman. New York: Simon & Schuster, 1955, cited in *Scientific Creationism*, ed. Henry Morris, Ph.D. San Diego, California: Creation-Life Publishers, 1974.
6. Lewis, C. S. "Is Theology Poetry?" *Oxford Socratic Club Digest*. 1944, cited by Morris, Henry in *That Their Words May Be Used Against Them*, 404–05.
7. Brown, Walt. *In the Beginning: Compelling Evidence for Creation and the Flood*. 6th ed. Phoenix, Arizona: Center for Scientific Creation, 1995, 3.
8. Lyell, Charles. *Life, Letters and Journal, I*. 253, cited by Immanuel Velikovsky in *Earth in Upheaval*. Garden City, New York: Doubleday & Company, Inc., 1955, 234–35.
9. Lyell, Charles. cited by Stephen Jay Gould in "First Man of the Earth." *Nature*, vol. 352 (August 15, 1991), 577–78.
10. Hitching, 226–27.
11. Lyell, Sir Charles. *Principles of Geology, 12th ed.* 1875, 318, cited by Velikovsky, 27.
12. Darwin, Charles. *The Origin of Species*.
13. Darwin, Charles. "Autobiography," reprinted in *The Voyage of Charles Darwin*, edited by Christopher Rawlings (BBC 1978), "A Scientist's Thoughts on Religion," *New Scientist*, vol. 104 (December 20/27, 1984): 75, cited in Morris, 357.
14. Ibid.
15. Ibid.
16. Denton, Michael. *Evolution: A Theory in Crisis*. Bethesda, Maryland: Adler & Adler, 1985, 66.
17. Huxley, Aldous. "Confessions of a Professed Atheist." *Report: Perspectives on the News, Vol. 3* (June 1966): 19, cited in Morris, 447.

18 *Humanist Manifesto II*, ed. Paul Kurtz. Buffalo, New York: Prometheus Books, 1973, 15–18.
19 Bozarth, G. Richard. "The Meaning of Evolution." *American Atheist* (February 1978): 30, cited in Morris, 375.
20 Denton, 67.

Chapter 3

THE SCIENTIFIC AGE OF THE EARTH AND UNIVERSE

When I get a chance to talk to students and young people, one of the questions I often ask is "How old is the earth?" I have stopped being surprised at the range of answers I get. Some students give an answer that involves a variation of "millions of years." Most say "billions," though few seem to have any concept of how many billions. Some have even told me "infinity." Yet one answer that I have seldom gotten is "six thousand years." Why is that? Until the advent of the so-called scientific age, "six thousand years" would have been the answer almost universally accepted by most biblical scholars. Does "scientific" evidence really supersede biblical history? If the universe is really only six thousand years old, should there not be some evidence that can be interpreted to support such a fact? It may not be feasible to scientifically *prove* that the world is only six thousand years old, but neither is it possible that "science" can establish that the earth is billions of years old. In fact, as with the wine created by Jesus at the wedding in Cana, scientific evidence is not even a plausible way to determine the age of a supernaturally *created* object.

There are four things that I hope to establish in this chapter: (1) There are philosophical roots to the idea of a vast age for the earth and solar system that are demonstrably antibiblical. The concept of billions of years of time is irreconcilable with the biblical account of creation and history. In fact, the concept did not even arise with, nor is it required by, "scientific" evidence. (2) Since "science" requires interpretation of data, which often depends largely on presuppositions, scientific methodology alone can never determine the true age of the solar system. (3) I want to give you some guidelines that will help you sort through varying interpretations of data, and help you select the interpretation that is the best and most straightforward fit. (4) I want to give you some pieces of data that can clearly be interpreted to indicate that our earth and solar system cannot be more than ten thousand years old. That is what I hope to accomplish in this chapter. However, if I accomplish nothing more than instilling the seeds of doubt in your mind as to the adequacies of the scientific method and the idea of billions of years of time, I will consider this chapter to be mildly successful.

(1) Many modern scientists seem to be oblivious of, or to intentionally ignore, the history of the current concept of a vast age for our planet and solar system. However, understanding the roots of such an idea is absolutely crucial to establishing its veracity and trustworthiness. Charles Lyell was a geologist who, in the early nineteenth century, promoted the idea of uniformitarianism. Interestingly, Charles Darwin, author of *The Origin of Species*, not only greatly admired Lyell, but used Lyell's precepts as foundational and essential to his own theory of evolution. The principle of uniformitarianism basically states: "Present continuity implies the improbability of past catastrophism and violence of change, either in the lifeless or the living world; moreover, we seek to interpret changes and laws of past time through those we observe at the present time."[1] In other words, no process took place in the past that is not taking place at present, and at essentially the same rate (consequently, the possibility of a worldwide flood has been philosophically eliminated even from discussion).

Darwin tied his concept of evolution inextricably to Lyell's concept of uniformitarianism. As H. F. Osborn, an early evolutionist who wrote in support of Darwin, stated, "This [idea of vast eras of time uninterrupted by catastrophes] was Darwin's secret, learned from Lyell."[2] However, such a postulate is not "scientific;" it is a philosophy of naturalism. And, therefore, any conclusions reached can be no truer than the initial presupposition.

In 1841, Lyell visited Niagara Falls in order to promote his theory of geologic uniformitarianism. His purpose was to find geologic features that would take longer than the biblical account of history would ostensibly allow. (Note: this, in itself, demonstrates that at that time, there was not much disagreement over what the Bible had to say on the subject. If the biblical account had been so unclear, or subject to such varying interpretations, there would have been no need to prove it wrong. There most assuredly is a spiritual root to the very idea of vast ages of time; and it is a root that exists in total opposition to the scriptural record.) Lyell estimated that the falls eroded at the rate of about one foot per year. Since the falls were thirty-five thousand feet from the lake into which they fed, he concluded by this one feature alone, biblical chronology was overturned. (Note: biblical teachers of the time, with little or no knowledge of geology, felt pressed by such "geological ages" to allow more time in the Bible. Thus was born the "day-age" theory that claims each day of creation to be a vast era of time. Realize there is no intrinsic hermeneutical reason for this. See also chapter "The Biblical Account of Creation.") Even the residents of the area at the time argued that the falls were known to erode an average of three to four feet per year.[3] Current estimates of four to five feet per year (and thirty-five thousand feet) would put the maximum time at closer to seven thousand years assuming no earthquakes sped up the process, the falls started at the lake (highly unlikely), the rate has been constant, no cataclysmic event (like the flood) was responsible, and so forth. According to Lyell: "All theories are rejected which involve the assumption of sudden and violent catastrophes and revolutions of the whole earth, and its inhabitants ..."[4] That is clearly not a principle of science; it is a statute of belief! And it is a thinly veiled—but absolute—rejection of the possibility of the global Flood of Noah. Nevertheless, the philosophy of uniformitarianism became the foundation of all geology—and the root of evolutionary theory. Which necessarily means that, if there was a worldwide cataclysmic flood, we would not be able to find out from Lyell's geology, since his geology began with the assumption that such an event could never have taken place. Evidence to the contrary had to be reinterpreted or ignored.

Not so incidentally, such reinterpretation of the evidence of geology fueled the invention of the Ice Age Theory (see chapter "The Myth of the Ice Age").

Charles Lyell and Charles Darwin did not invent the idea of vast eras of time—the concept had existed in pagan philosophies in different forms for numerous centuries before them. But the root of its acceptance in modern society can be clearly traced to a combination of their writings. Lord Kelvin was a Christian physicist (after whom the absolute temperature scale was named). He once calculated that if the earth had begun in a molten state, as hot as theoretically possible, using the known properties of the earth and rates of cooling known to physics, it would take a maximum of fifteen to twenty million years for the earth to reach its current temperature. Kelvin did not say that the earth had to be that old or even that the earth could be that old; he only said, by the laws of physics, it could not be older than that. For that reason, Darwin referred to Lord Kelvin as an "odious specter." That may strike even the casual observer as a somewhat less than scientific analysis of Lord Kelvin's scientific abilities or the validity of his conclusions—not exactly something that might be referred to as adequate "peer review." Indeed, Darwin said in his writings, "I am greatly troubled at the short duration of the world according to Lord Kelvin, for I require for my theoretical views a very long time."[5]

Inherent in Darwin's statement, are two things of note: (a) Kelvin was not trying to prove or promote any particular belief; he was just trying to interpret the evidence wherever it led. On the other hand, (b) Darwin was determined to oppose or eliminate any information that undermined his theory. He was greatly troubled by the real evidence. Such a statement should never be construed as "science." In fact, it should be incredible to anyone who has studied history that Darwin is considered a "scientist" and Christian physicists like Lord Kelvin and Sir Isaac Newton are generally ignored when they speak of origins. In any case, the essential point is that the concept of vast eras of time was the foundational undergirding to the idea of evolution, because it was more than obvious that nothing like Darwin's concept of slow, gradual change could have occurred in only six thousand years. Prominent evolutionist Richard Dawkins stated that "Darwin's own bulldog, Huxley, ... warned him against his insistent gradualism, but Darwin had good reason. His theory was largely *aimed at replacing creationism* as an explanation of how living complexity could arise out of simplicity. Complexity cannot spring up in a single stroke of chance: that would be like hitting upon the combination number that opens a bank vault [in one blind attempt]. But a whole series of tiny chance steps, if non-randomly selected, can build up almost limitless complexity of adaptation. It is as though the vault's door were to open another chink every time the number on the dials moved a little closer to the winning number. Gradualism is of the essence. In the context of the fight against creationism, gradualism is more or less synonymous with evolution itself. If you throw out gradualism you throw out the very thing that makes evolution more plausible than creation [italics added]."[6]

In other words, there are two absolutely essential principles to consider in looking at the history of uniformitarianism (used here as a synonym for the idea of a universe billions of years old) and evolution: (a) early evolutionists considered their theory as a replacement for, not a type of, creationism. Evolution theory and creationism are necessarily diametrically opposed because evolution theory was begun as an attempt to eradicate creation as an option, and (b) the element of vast eras of time is the foundational requirement for evolution theory. It is a requirement of evolution theory, but not of the laws or finds of science. If the concept of billions of years of time is eliminated, evolution theory dies with it.

(2) The second area to address centers on interpretation of evidence and beliefs. Recall that if the earth formed by natural process, science might be able to offer some reasonable hypotheses to explain both that formation process and the duration of same. However,

none of the "natural processes" seem to be repeatable, and all natural explanations of planet (or star) formation fall woefully short at best. Furthermore, if the earth did not form by natural processes, but by divine fiat, then scientific explanations are meaningless. At best, science can look at some current processes, but when it has to extrapolate back in time, it must make assumptions. The validity of the assumptions often depends on underlying beliefs and usually cannot be tested. In addition, interpretation of observational data usually needs to be done in conjunction with all other such data. For example, one partial fossil does not constitute proof of a missing link!

To illustrate the difficulties in taking current process rates and utilizing them to make definitive conclusions about past origins, let us look at some simplified examples. Suppose you entered a room and found three processes occurring simultaneously. From those, assuming all processes began at the same time, we will endeavor to "scientifically" calculate the "age" of the room.

(A) A candle is burning on the counter. It is measured to be five inches tall. After observations, it is determined that the current rate of melting appears to be one-half inch per hour. From those simple pieces of information it should be evident that one could readily estimate the remaining life of the candle. However, it should be equally apparent that determining how long the candle has been burning is a far more complex question with no conclusive answers. In fact, without the answer to some essential questions, it may be that no answer is even possible. However, for the sake of the illustration, let us look at some of the questions to be addressed: (a) How tall was the candle to begin with? (can we assume that the maximum would be less than the height of the room?) (b) Was there more oxygen in the room previously that could have increased melting rate? (c) Has the candle been burning continuously or could it have been stopped and restarted? (d) Was the candle made of consistent material and was it consistent in diameter throughout its height? (Note: if the candle leaves a residue as it burns, then some of this may be measured or guessed at. But if the candle turns to oxygen as it melts, which is indistinguishable from the other oxygen in the room—a situation comparable to most radioactive decay by-products—then the determination of residual oxygen becomes virtually impossible.) (e) Was the wick made of a consistent material? And, of course, there are probably numerous other questions that would need to be addressed even in this simplified example.

In this case, if the assumption is made that the candle was consistent in material and wick throughout its height, and that it was never put out and restarted, and that its maximum initial height was the height of the room, then the best that could be arrived at is a very rough approximation of the maximum age of the candle. In this case, assume that the height of the ceiling above the placement of the candle is six feet. That yields a maximum initial height of the candle of seventy-two inches melting at half an inch per hour for a maximum burning time of 144 hours, or six days. Anything more than that would presumably require some kind of intervention or unverifiable change.

(B) A pot of water is boiling on the stove. The container holds a maximum of two quarts, and currently contains one quart (thirty-two ounces per quart). Measurement indicates that at current rates, eight ounces of water evaporate per hour. Once again, it should be fairly simple to determine how long this boiling process could continue (realizing that as the water level diminishes, the rate of evaporation would be expected to increase.), but far more difficult to estimate how long it has been going on. There are simply too many variables that must remain unknown and unknowable. For example: Has the fire on the stove been constant? How much water was in the pot in the first place? What was the temperature of the water when first placed in the pot? Could it have been ice? Could any water have been added to the pot at any time? Was the stove ever turned off and restarted? How does the amount of water in the pot determine evaporation rates? Undoubtedly, there are many more possible questions and many variations of the questions addressed. Again, we can make some "reasonable" assumptions, but the best it will give us is a maximum time period (much like Kelvin's maximum estimate for the time of the earth's cooling that so troubled Darwin) for the boiling process to have gone on. Assume that the pot of water began at room temperature and that it was full to begin with. Once boiling, if eight ounces evaporate per hour, then it would have taken four hours to lose one quart of water. If one wanted to take into consideration the length of time that it would take two quarts of water to reach boiling point, then perhaps some other experiments could be done. In addition, one might assume that evaporation rates could increase slightly as the water level diminishes (since the same amount of heat is being applied to less and less water). However, it should be clear that, at most, only a small amount of time would be added to the overall estimate. Instead of four hours, a scientist might, therefore, postulate a maximum duration for the past boiling of six hours.

(C) Finally, there is a battery-operated clock on the wall that indicates the current time. Measurements indicate the battery has a current available potential of 1.3 volts. The rate at which it is losing potential is measured to be .05 volts per month. The battery can probably keep the clock running until it has .8 volts remaining. As in each earlier problem, it is again clear that estimating the remaining life of the battery is much easier than determining the previous life of the battery. Again there are questions that may be impossible to answer: Was the battery fully charged to begin with? What is the maximum charge for the battery? Does the decay rate depend on the age of the battery? Does the battery potential decay linearly (most likely not, and we probably cannot determine what the decay rate would have been at any point of the past unless we were there to test it at each stage)? Could room conditions have varied and could such variations affect battery decay rates (such as much colder temperatures slowing the decay)? Once again the scientist is faced with making assumptions (which may or may not seem reasonable, depending on one's presuppositions, but which,

The Scientific Age of the Earth and Universe

in the end, could at best produce a rough guesstimate of maximum age and nothing else). In this case, assume that the maximum potential of a new battery is 1.6 volts (similar to your standard flashlight battery) and that decay can be approximated as linear. Since .3 volts have been lost at the rate of .05 volts per month, the process could have been going on for a maximum of six months.

Clearly, the three different sets of observations and calculations yielded very different results: The candle could have been burning for up to six days, the pot boiling for up to six hours, and the clock running for up to six months. So now the question becomes, if all three processes are assumed to have begun at the same time, which would be the best estimate for the age of the room?

Once again, the "scientist" would have to make some untestable assumptions. For the purpose of this discussion, since the scientist is trying to determine the age of the room, it will have to be assumed that all three processes began at roughly the same time, and all three have been unaffected by outside influences. In that case, it should be clear that the estimate for the age of the room must be the process that was going on for the shortest length of time. The clock might have been running for six months, but without historical data on the battery it is also possible that it has only been running for six hours. The candle might have been burning for up to six days, but it is possible, and even likely, that it was only burning for six hours. On the other hand, the pot of water, which had a maximum boiling time of six hours, could not have been boiling for six days. The water would have been gone and the pot destroyed long before six days had passed. That means, other things being equal, it is not the process that could have been going on the longest that yields the most likely age of a system, but the process that could have been going on for the shortest length of time. The youngest age is the one that would most likely determine the upper limit for the age of a system, and it also often involves the least amount of speculation or other variables. It is also often the one that is most observable and testable.

So, though some scientists like to point at starlight that appears to have come from stars that are billions of light-years away as "proof" that billions of years of time have occurred, they are overlooking or intentionally ignoring a wealth of processes that simply could not have been going on for more than thousands of years. And the idea or belief that stars are billions of light-years away, or that light from them took billions of years to arrive at earth, is quite likely the least testable of all such hypotheses. And it is clearly overruled by not only a large number of measurable processes, but by the revealed history recorded in God's Revelation of earth's history—the Bible.

There is virtually no such thing as a scientific observation that does not require interpretation. And there is no such thing as an interpretation that is completely objective. As Stephen Jay Gould, professor of geology and paleontology at Harvard University, admitted, "Facts do not 'speak for themselves'; they are read in the light of theory. Creative thought, in science as much as in the arts, is the motor of changing opinion. Science is a quintessentially human activity, not a mechanized, robot-like accumulation of objective information, leading by laws of logic to inescapable interpretation."[7] (So in reality "science" is not the factual interpretation of accumulating data, but the creative speculation of changing opinion! Who knew?) Most laymen, and unfortunately, many scientists, do not seem to realize how much subjectivity and philosophy have colored interpretations of data—especially those that involve dating methods. And, in general, the older the date, the less objective is the interpretation—if for no other reason than the fact that the available data are less testable and more speculative.

(3) If scientific theories or explanations do depend on subjective interpretations of data, there should be some criteria that would enable one to choose the more reasonable "scientific" explanation. Sometimes, as nonscientists, we might find it difficult to decide between varying interpretations of the same evidence. For example, evolutionists (or long-

The Scientific Age of the Earth and Universe

age creationists) can look at the ice layers of Greenland and say, "At current rates, this had to take hundreds of thousands of years to build up; therefore the earth is older than hundreds of thousands of years." Then they will examine the amount of dust on the moon and realize, at current rates of accumulation, it would take less than ten thousand years to achieve current levels. They then conclude that rates were much slower in the past. (Note: this is not based on sound principles of science, but most often on a presupposition. Therefore, conclusions are only true if the presuppositions were true.)

On the other hand, the young-earth creationists look at exactly the same data and conclude that the moon and earth system is less than ten thousand years old. When they look at the amount of dust on the surface of the moon, they conclude that, at current rates, this could have been accomplished in only a few thousand years and is consistent with the biblical chronology. Yet when they look at the ice layers in Greenland, they conclude that rates of ice-accumulation were much more rapid at some time in the past (i.e.; following the flood). The data were the same for both sets of observers, but they arrived at two very different sets of interpretations. Consequently it is legitimate to ask how one might decide which set of interpretations is best.

I would suggest that there are three tests to apply: (a) Which interpretation fits the data with the least amount of manipulations? (For example, the fossil record—and its non-existent transitional links—requires constant manipulation to fit the idea of evolution, but virtually none to support the biblical account of creation and a global flood.); (b) Which interpretation fits without postulating events that are different from, or contrary to, the direction that such processes normally or naturally take? (For example, as far as the rate of dust buildup on the moon is concerned, we would expect, according to observations and the known laws of physics, the accumulation rate of dust to have been greater in the past rather than less. No known mechanism can explain a recent increase in dust in the solar system: the amount of available dust should be diminishing as it is gravitationally swept up by sun, moon, and planets. Scientists can postulate as many up-and-down fluctuations as they like, but it is not what one would expect to see, and it was never predicted by the laws of science, so it should not be considered the best scientific interpretation of the available data.); (c) Recorded history or Revelation! If, for instance, God revealed to mankind how old the earth was, then it would be far easier to interpret the physical data that is available. For example, suppose that one hundred years in the future, scientists unearth a previously unremembered car from the twentieth century. They try to determine its age from its conditions and features. They compare two features: its headlights and its sound system. In this instance, suppose the car has halogen headlight bulbs and an 8-track tape system. Consequently, one scientist determines that the car must have been from the late 1980s or later, because cars were not equipped with halogen bulbs before then. He concludes that someone must have retrofitted the car with an 8-track tape deck because they liked the sound. Alternatively, the second scientist decides the car must be from the early 1960s because no one manufactured a car with an 8-track tape system after that. He concludes the owner must have later fitted a new headlight bulb system in an older car. Now, which is correct? There may be no way to tell, though there may be some principles that could tend to sway us in one direction (for instance, it might be considered more likely that someone fitted new headlights to an old car than that they fitted an old, outmoded sound system to a new car). However, if the future scientists happened to look in the glove compartment and find the original bill of sale, complete with serial number, dated at 1965, they would know which interpretation was correct. They had historical revelation.

In lieu of the foregoing, I would suggest that the third "test" far outweighs in significance either of the first two tests. If the history of the creation of dinosaurs (for instance) and the time of their demise was historically recorded, it would clearly make the interpretation of the fossil record that much easier. That, indeed, is the situation in which mankind finds itself. The truth about origins and time can be known rather than based on imaginative speculation. It does not have to remain an enigma or a guessing game. As a matter of fact, historically speaking, mankind was given the bill of sale to the earth and universe. The serial number is attached, the date is in writing, and the author is available to us for consultation (see chapter "The Biblical Account of Creation").

(4) The fourth and final area to examine is scientific evidence that can be interpreted to indicate an earth or a universe consistent in age with the biblical account of history—namely about six thousand years. The presentation of material in this section is decidedly autobiographical, since it is also, in part, my journey.

The moon is 1/81st the mass of the earth. In fact, it is the largest known satellite in our solar system in terms of its percentage of the parent planet. It is also in an orbit that is almost a perfect circle. If its orbit could be reduced to scale to fit on a piece of notebook paper, to the average human eye it would be visually indistinguishable from a circle made by a compass. To the physicist, that means that the moon is too large a satellite to have been gravitationally captured by the earth in its present orbit. Consequently, most scientists of any persuasion believe the earth and the moon must be the same age: they must somehow have formed as a pair. One of the goals, then, of the lunar missions was to demonstrate that the moon was the same age as the earth—4.6 billion years old.

But in 1969 scientists and NASA were faced with a dilemma. I remember watching a morning newscast a day or two before the first lunar module was scheduled to land on the moon. They were interviewing a scientist who was warning of the tremendous danger of landing on the moon. He said, as I recall, "I will stake my job and reputation on the fact that the lunar module will sink into dust from which it will be unable to extricate itself." In fact, before instruments were sent to the moon to collect data and make readings, evolutionist Isaac Asimov made similar dire predictions of overwhelming levels of dust on the moon's surface: "I get a picture ... of the first spaceship, picking out a nice level place for landing purposes, coming in slowly downward tail-first and sinking majestically out of sight."[8] I had a science fiction record album at the time that presented a similar scenario. In it, the astronauts slid into a crater and disappeared into the depths of dust that had accumulated in the crater. One reason that a vast amount of dust was hypothesized to be infilling craters was that there was very little evidence for its existence on the surface, even though it was believed to exist in large quantities. By the time of the lunar landing, manned missions had been around the moon, photographed it extensively, bounced laser beams off it, done spectrographic analyses, and found no evidence of a massive layer of dust. And yet predictions based on the belief in a solar system that was billions of years old insisted on deep layers of dust on the moon! Measurements on earth have indicated that the earth accumulates about fourteen million tons of cosmic dust per year as it sweeps through the solar system. Now, that may not be a lot spread over the entire surface of the earth; but the moon has no atmosphere, no wind, no rain, no water, and virtually no erosion other than gravitational. The dust would simply build up. (For example, on a straightforward linear model, .0001 inch of dust per year would result in a uniform layer of dust one inch deep in ten thousand years; but in 4.6 billion years that same rate of accumulation would result in a layer over seven miles deep—enough to cover every high mountain of the earth and then some!)

I remember watching in some anticipation as the lunar module made its descent toward the moon's surface. The cameras could not be turned on because of the heat and glare from the retro rockets, so NASA provided a cartoon-like simulation. It showed a crater of dust being formed around the lunar module approximately three to four feet high. The equipment and astronauts were prepared to deal with up to several feet of dust. Yet a few minutes later, news narrator Walter Cronkite announced that NASA control had informed them that there was a mistake in the simulation and no crater of dust had been formed. In fact, there was only an inch or so of dust on the moon.

Dust on the moon—or, the lack thereof. [Fig. 31, *In the Beginning: Compelling Evidence for Creation and the Flood*, 8th ed., 2008, by Walt Brown, p. 38, from composition by Bradley W. Anderson using NASA photos.]

Yet, in spite of measurements and observations to the contrary, equipment was still designed to deal with at least several feet of dust. Several feet of dust that was not there. So, why is it not there? It has to be there—meteoric influx alone would produce far more surface dust in 4.6 billion years than is currently accounted for.

Most scientists now say, in retrospect, that the rate of influx of lunar dust was less in the past. But what is that conclusion based upon? Remember the criteria I suggested for deciding between two alternative explanations? So ask yourself, as the planets and moons sweep through our solar system and gravitationally vacuum up dust, does it seem more likely that the amount of available cosmic dust would be increasing with time or decreasing? I would suggest that unless some mechanism is discovered that is creating dust in the solar system (maybe we are the dumping grounds for all the alien vacuum cleaners in the universe?), it is far more likely that the current rate of accumulation is less than it was in the past. Is it possible that there was less dust accumulating annually in the past? Maybe. But how could such a theory be tested? It is not based on observations, but on the absence of observations. Indeed, the theory that dust accumulation rates in the past were less than current rates exists at all only because it has to in order to maintain the age of the solar system at billions of years. And remember, no Cal Tech astronomer or NASA prognosticator predicted this, based on evidence, in 1969 before the lunar landing. If it was something that was expected from other natural processes, it should have been predictable. A poor scientific theory is one that predicts nothing, but explains everything after the fact. It is interesting to note that in the years following 1969, in spite of apparently insufficient dust to bolster the story of billions of years of lunar wandering through the solar system, no NASA scientist publicly advocated reducing the age of the moon as an explanation of the missing dust (perhaps the most logical conclusion to be drawn from the data). Instead, they frantically reworked the data itself. Now, after nearly forty years of manipulation, NASA claims that its initial estimates on rates of dust-accumulation were incorrect. They now insist that the actual rate of build-up is consistent with an age estimate of billions of years. Maybe so. Personally, I doubt it. For the last forty years, instead of altering either their theories to fit the data, or even their conclusions drawn from the data, they have worked instead to *change the data*. It sounds suspiciously like the shoemaker who, when confronted by the fact that the shoe he made with his own measurements is too small to fit its owner, tells the owner that his feet must have grown and he needs to amputate his toes. Maybe it is true that the data was wrong in 1969, but the question still remains: why did no one at NASA use the data that did exist in 1969 to establish the age of the moon at less than 10,000 years (which was seemingly the most straightforward interpretation of the data)?

A second physical indicator that could be used to demonstrate a solar system with an age of only thousands of years is the continued existence of comets. There are three major observations that can be made of comets: (a) Comets travel around the sun and diminish in size—they apparently melt; (b) Based on current rates of melting, most short-period comets must be quite young (on the order of thousands of years); (c) Comets do not appear to be being produced anywhere within our observable solar system. Now remember, observations and data need to be interpreted, but the interpretation depends largely upon presuppositions. I recall, in my astrophysics class, my professor admitting to some perplexity when trying to explain the apparent youthfulness of comets in a solar system that was ostensibly billions of years old.

In fact, according to my college text, *Introductory Astronomy and Astrophysics*: "Since comets quickly waste away on a geological time scale, the supply must be constantly replenished if we are to see them today. [Note: this is not the most clear and straightforward interpretation of the available data, it is an interpretation based on the belief that the solar system must be billions of years old.] It is presumed that cometary nuclei were formed with the Solar System about five billion years ago. To account for their continued existence, Jan Oort has hypothesized a spherical *comet cloud* [italics in original] between [30,000 and 100,000 Astronomical Units] from the Sun wherein reside about [100 billion] nuclei... Sad to say, few theories of cometary origin exist, and fewer still are even partially convincing."[9] Notice, once again, that the primary difficulty with the continued existence of comets lies in the presumed age of the solar system. Bizarre, untestable, and even unreasonable theories are proposed in an effort to explain the continued existence of comets for the sole purpose of allowing observations to fit the belief in a universe that is billions of years old.

In 1985, when Halley's comet was set to pass by earth once again, I went to a symposium on comets put on at the Jet Propulsion Lab (JPL) in Pasadena, California. During the question-and-answer phase, I posed the following question: "If it were not for your presupposition that the solar system is billions of years old, is there one single piece of objective data that supports the existence of the 'Oort cloud'?" After stumbling for a while, the lecturing astronomer finally (after a prompting from someone in the audience, "Can you just answer yes or no?") said, "No!" He then went on to explain why, in his belief system, it must be there anyhow. In reality, he demonstrated that when observations do not fit the strongly held belief in the concept of billions of years of time, it is the observations that are altered and fictional data that is invented. A straightforward and perfectly rational interpretation of the cometary evidence is that comets, many of which only survive for tens of thousands of years, are still visible in the solar system precisely because the solar system is only thousands of years old. Sadly, the nature of the opposition to such an interpretation far surpasses the rational nature of the evidence.

Admittedly, there are some observations (actually, primarily one) that could be logically interpreted to imply an old universe, but recall the examples from the second principle established earlier, that it is generally the youngest indicators that are most testable. If some indicators imply that the universe may be billions of years old, but others indicate that the universe could not be more than thousands of years old, logic compels that the upper limit must be placed by the youngest indicators. (Remember that in our example, the clock could have been running for several months, the candle could have been burning for an absolute maximum of five or six days—likely much less—but the pot could not have been boiling for much more than about six hours. Therefore, we would conclude that if all of those processes began around the same time, the best indicator for the age of the system would be the pot of water.)

When I was in college in the early seventies, I took some classes in astronomy, and in one of these we had to do a research paper. I found a topic that fascinated me because it demonstrated the very tenuous nature of the grip astronomers had on understanding even the closest "star," the sun. Astronomical journal headlines at the time referred to the "great neutrino mystery," or "the case of the missing neutrino." At the time, I did not realize that the controversy could be simply resolved if the theorized age of the sun was diminished from billions of years to thousands of years. Theory says that a young star is fairly homogenous throughout and as it gravitationally contracts, tremendous heat and pressure at its core begin a process of nuclear fusion: hydrogen gas is fused into heavier helium gas and tremendous energy is released. Over billions of years, the core of helium increases in size and nuclear fusion increases as well. In nuclear fusion, neutrinos—tiny packets of energy that travel at the speed of light—are released. Since the sun is estimated to be roughly five billion years old, astronomers believe they know how large the core of helium should be (the radius is currently estimated to be 175,000 kilometers and the density to be roughly fourteen times that of lead) and how many neutrinos are being released. And yet when physicists began actually trying to measure the influx of neutrinos from the sun, they found far fewer than they had predicted. The headlines read "Neutrinos Are Missing!" The missing neutrino dilemma led to no end of speculation in the physics and astronomy journals. Some scientists went so far as to speculate that perhaps the sun's furnace cycled off and on and it was currently in an off-cycle (no one had any reasonable idea of why that might be the case, or what physical mechanism could account for that). Of course, such a scenario was not a prediction of solar theory. It was an after-the-fact attempt to rationalize the available evidence into a theory for which the evidence was unavailable. Other scientists simply said that the measurements of neutrino numbers had to be wrong. Their belief system about the age of the sun and the processes going on inside were sacrosanct—untouchables! Such a belief system could not be altered by the data! Consequently, they argued that the methods of capturing and counting neutrinos needed to be refined. Yet, after several years of refining the methods and numerous recalculations, the neutrinos were still missing. Perhaps not so strangely, one explanation that was conspicuously absent from most American scientific journals was that the sun was less than ten thousand years old and nuclear fusion had not yet begun in significant measure.

Now, after decades of attempting to refine not only the methods of data accumulation, but the theories themselves, some scientists believe the data can be made to fit the expected nuclear fusion model (even if that were true it would not preclude the sun from having been created six thousand years ago to do precisely what it is doing today). Nevertheless, it should be clear that numerous convoluted and extraordinary explanations were attempted for many years in order to avoid the obvious explanation that the sun was not yet billions of years old.

However, in 1976, a team of Russian astronomers published research in the British scientific journal, *Nature*, correlating computer models with actual measurements of data from the sun. Their conclusion: nuclear reactions "are not responsible for energy generation in the Sun."[10] Their results showed the sun as "bearing the characteristics of a very young homogenous star that corresponds with the early stages of the computer models."[11] According to the article, there were several measured observations that were entirely consistent with a young star that was thousands of years old, but completely incompatible with any model of a star billions of years old. In fact, a leading solar astronomer, John Eddy, publicly stated in 1978 that there is no evidence based solely on solar observations that the sun is actually 4.5 billion years old. "'I suspect,' he said, 'that the sun is 4.5 billion years old. However, given some new and unexpected results to the contrary, and some time for frantic recalculation and theoretical readjustment, *I suspect we could live with Bishop Ussher's value for the age of the earth and sun.* [Bishop Ussher was a brilliant Hebrew and biblical scholar who used the chronologies and genealogies of Scripture to calculate the date for the creation of the universe to have been 4004 B.C.] *I don't think we have much in the way of observational evidence in astronomy to conflict with that* [italics in Lubenow].'"[12] According to Eddy, astronomers and astrophysicists rely on paleontology (essentially the dating of the fossils by evolutionists) for determining the age of the sun. "Astronomy, as an observational science, can say nothing about chronology as far back as [4.7 billion] years."[13] Such a statement by one of the world's foremost astronomers should have been a front-page, earth-shattering headline: "Observations [data] cannot tell us that the solar system is billions of years old, we must rely on the theory of evolution ("paleontology") for our chronology of the solar system." And, indeed, even according to Eddy, the data that are available could be used to bolster a belief in the biblical chronology established by Archbishop Ussher, or a sun and solar system that is about six thousand years old. But that possibility would require a Creator and thus is rejected as unscientific.

One of the things that I quickly and consistently noticed in my astronomy and astrophysics classes was that clear and convincing evidence for a youthful universe was never interpreted as such. Complicated and convoluted explanations of visible phenomena were often invented for the sole reason of extending the age of the universe or solar system. For example, spiral arm galaxies appear to be revolving. That is the self-evident nature of their appearance. Most of the mass (ninety-eight percent or more) appears to be concentrated at the center of the galaxy, and the "arms" have every appearance of winding themselves around the core as would be predicted by the laws of physics. And yet if these galaxies have been around for ten billion years or so, the "spiral arms" should have completely wound themselves around the core within the first revolution. Many convoluted and entirely unwarranted explanations have consequently been put forth—such as trying to explain the "arms" of stars as density waves of energy that temporarily bunch the stars together—but all such hypotheses remain outside the realm of testing, and the only reason for postulating such a thing in the first place is to circumvent any explanation that eliminates billions of years from the universe.

Astronomers have also observed many clusters of galaxies that should not still exist. From their estimates of the masses available, they conclude that there is not enough material to keep these galaxies gravitationally bound over billions of years. Yet the galaxies are together in observed clusters. Since the galaxies appear to be bound together in a universe that is supposedly expanding, scientists have speculated on the existence of massive black holes at the center of each galaxy, or "dark matter," or antimatter, and so on. None of those complicated hypotheses were initially suggested by the data, but they are essential in order to explain how galaxy clusters can still be together when there is not enough visible mass to hold them in a cluster for billions of years of time.

"Quasars" are another interesting phenomenon because they also create logistical nightmares for the concept of billions of years of time. Quasars, or "quasi-stellar objects" appear to be quite normal stars, but some of them display very large red shifts of the light coming from them.[14] The red shift is usually interpreted to be the result of the Doppler effect. The larger the red shift, the faster the object is estimated to be moving away from the earth. All such motion away from earth is interpreted to be a product of the big bang and the beginning of the universe. Therefore, the faster an object is moving, the farther away it must now be. Some quasars display a very large red shift and consequently are estimated to be moving away from earth at better than ninety percent of the speed of light. Of course, if some quasars have been doing that for billions of years, they should long ago have disappeared from view. The fact that they are still visible leads some scientists to conclude that they are giving out tremendous amounts of energy. In fact, some calculations show a single quasar producing more energy than the entire Milky Way Galaxy of one hundred billion stars combined. This has led to all sorts of speculation and nonsensical explanations. Some astronomers guess that a quasar is the "doorway" to another universe; all the energy of that universe is theoretically being sucked into a massive black hole and given off in our universe as a "white hole." Of course, once again, the only reason for such ridiculous speculation is an effort to cling to billions of years of time where they do not seem to be available. If the universe is only on the order of six to ten thousand years old, even if the observed red shift from quasars is due to relative motion (and there is no way to establish that), there is no problem with any of the observations.

In addition, there are alternative possibilities to explain the red shift that may not be due to the Doppler effect at all. Recent studies have concluded that the speed of light may be slowing down—indeed, some studies have indicated it could have been dramatically higher in the past. If so, the red shift we see may be related to the slowing of light and not the movement of the star. Studies have also indicated that red shifts occur in quantum amounts (For example, it would be like measuring the speed of cars moving away from your position on the freeway and discovering that they were moving in multiples of ten miles per hour: ten miles per hour, twenty miles per hour, thirty miles per hour, and so on), which may further indicate the shifts are not Doppler related. Of course, if red shifts of the light from distant galaxies are not related to relative motion, the entire expanding-universe scenario could turn out to be fiction as well. It is certainly based on untestable assumptions. And, of course, the entire foundation of the "big bang" hypothesis for the beginning of the universe depends upon the current expansion of the universe, which in turn depends upon the Doppler effect interpretation of red shifts. And the truth is that many astronomy and physics texts are subtly dishonest in their portrayal of the data and its interpretation. For example, Holt's physics textbook for high school and college students states, "As scientists began to study other galaxies with spectroscopy, the results were astonishing: *nearly all* of the galaxies that were observed exhibited a red shift, which suggested that they were moving away from Earth. If *all* galaxies are moving away from Earth, the universe must be expanding [emphases added]."[15] The astute reader might notice that "nearly all" galaxies exhibit a red shift, but *not all*. That implies that some exhibit a blue shift. If interpreted to be due to the same Doppler Effect, then some galaxies must be moving toward Earth. The authors do not specifically mention that fact. Nor do they discuss how an expanding universe might have some galaxies that are moving in reverse. Did those blue-shifted galaxies hit an invisible wall in the cosmos? How did they bounce back? Why did they bounce back? The authors subtly transition to the statement that if *all* galaxies exhibited a red shift, it would imply that the universe was expanding. And maybe that is even true (though certainly untestable since it cannot even be demonstrated that the perceived red shift is due to the Doppler effect). But, it is also true, as they have previously only hinted at, that all galaxies do not exhibit a red shift. By their own analysis, if they were honest, they would be forced to conclude that the universe is not expanding. Instead, the authors make a quantum leap forward to state that the (now generally accepted) "expansion of the universe suggests that at some point in the past, the universe must have been confined to a point of infinite density. The eruption of the universe is often referred to as the *big bang*, which is generally considered to have occurred between 10 billion and 20 billion years ago [italics in original]."[16] An honest presentation of the data would at least have wrestled with the fact that some galaxies exhibit a blue shift and this alone calls into serious question the expansion of the universe and, consequently, the entire big bang theory. An honest portrayal of the data would also admit that the discrepancy between ten billion and twenty billion years is enormous and should hardly inspire confidence in the theory.

There are numerous other evidences that could be examined on earth that seem to indicate, at face value, a very young earth (such as the rapid decay of the earth's magnetic field strength, the erosion of the continents, carbon-14 buildup in the atmosphere, helium amounts in the atmosphere, chemical content of the oceans, and so on). But one very intriguing field that seems inexplicable by any "billions-of-years" scenarios is the existence of specific radiohalos in certain of the earth's "pre-Cambrian" granite rocks (these are the "foundational" rocks of the earth that contain no fossils). These radioactive halos, or "radiohalos," are microscopic discolorations produced by the radioactive decay of certain elements trapped within the rocks. (See photographs below.)

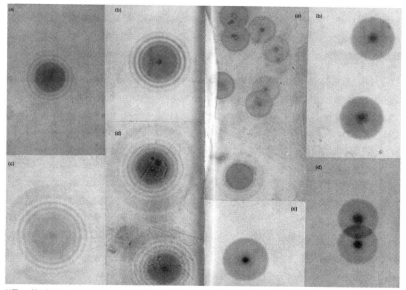

"**Radiohalos" photographed in granite rocks.** [Photograph from Gentry, Robert. *Creation's Tiny Mystery*, 3rd ed. Knoxville, Tennessee: Earth Science Associates, 1992, photo pages 1–2.]

Robert Gentry (considered by most to be the world's foremost expert in the field) has found that some radiohalos have been produced exclusively by the decay of Polonium-218. However, since Polonium-218 decays very rapidly, having a half-life of only three minutes, it does not exist in nature except as a very short-lived daughter product of uranium decay. Yet radioactive halos cannot form in a granite rock unless the rock is fully solidified. Therefore, unless the granite rock formed almost instantly, any polonium-218 halos found in these granite rocks should always be included exclusively within the halos of uranium decay. Dr. Gentry and others have conducted exhaustive experiments to try and account for the inclusion of polonium in any way and at any time other than at the initial formation of the rock. All such tests have thoroughly demonstrated—to date—the impossibility of the Polonium-218 being separately captured after the solidification of the rocks.

Dr. Gentry's interpretation is that these "pre-Cambrian" granites were formed instantly and the Polonium-218 was there at its inception ("primordial polonium"). This is problematic—to say the least—for current evolution theory, which claims that these rocks formed by the slow and gradual cooling of molten magma over millions of years.

In the last thirty years no credible, alternative interpretation has been offered that has not already been clearly falsified by Dr. Gentry's ongoing experiments. In fact, Dr. Gentry has proposed experiments that, if successful, could falsify his interpretation of the data (supposedly the hallmark of a "good" theory). Most evolutionary texts and publications continue to assert—with no evidence or observations—that these granite rocks formed by gradual cooling over millions of years. And if Dr. Gentry's observations and experimental data are mentioned at all, they are usually glossed over as a "tiny inexplicable mystery." Some scientists continue to insist that eventually a mechanism will be discovered to explain how Po-218 could have been incorporated into the granite rock millions of years after the rock originally formed. Nevertheless, it should be noted that wishful thinking does not generally qualify—or should not—as scientific interpretation of observations. In fact, this "wishful thinking," thinly disguised as "science," appears to be groundless speculation based solely on the fear of acceptance of the only reasonable alternative; namely that the earth is quite young and was very likely "created" in a very short period of time. Dr. Gentry refers to these tiny, spherical, Polonium-218 halos as "the fingerprints of creation."[17]

Notice that all of the situations and observations presented in this chapter have one thing in common: they are very difficult to explain, and they require complicated manipulations of physics that would never be predicted unless one begins with the premise that the universe is billions of years old and was formed by natural processes. None of the observations contains any inexplicable dilemmas or contradictions of physical laws if the solar system and the universe are only thousands of years old and were formed by the hand of the Creator. In fact, in each case, the more logical and straightforward interpretation of the data is that the solar system is quite young: Quasars may be normal stars moving rapidly away from us—we still see them simply because they have not been moving away from us for very long. (In fact, the "red shift" we see may be caused by something other than the Doppler effect and relative motion, such as the slowing of the speed of light.) Neutrinos are missing from the sun because it is quite young and has not yet begun massive internal nuclear fusion. It is shrinking because of gravitational contraction. Spiral arm galaxies have not wound themselves up because they have not been around long enough. Clusters of galaxies exist because not enough time has passed for them to have drifted apart. Comets are still visible because our solar system is quite young. The moon has only one to two inches of dust on its surface because the dust has only been accumulating for thousands of years. Polonium-218 halos exist in certain granite rocks (completely separate from any uranium halos) because the rocks were created in an instant with the polonium already present.

D. M. S. Watson, a British biologist, stated that "[Evolution is] a theory universally accepted, not because it can be proved by logically coherent evidence to be true, but because the only alternative, special creation, is clearly incredible."[18] The same can be said for the concept of billions of years of time. Billions of years are neither a problem nor a necessity for creation, but they are an absolute requirement of evolution theory. If, indeed, the solar system is only thousands of years old, then there is no possibility that it came about by natural processes in that length of time. Indeed, it was formed by the hand and the Word of the Creator. And that is precisely why there is such insistence on establishing the age of the universe in the billions of years. Because the only viable alternative is creation at the hand of the Creator.

Endnotes

1 Osborn, H. F. *The Origin and Evolution of Life*. 1917, 24, cited in Velikovsky, Immanuel. *The Earth in Upheaval*. Garden City, New York: Doubleday & Company, Inc., 1955, 24.

2 Ibid.
3 Velikovsky, 161–162.
4 Lyell, Charles, *Principles of Geology*, 12th ed. 1875, 318, cited in Velikovsky, 27–28.
5 Darwin, Charles, cited in Macbeth, Norman. *Darwin Retried: An Appeal to Reason*. Boston, Massachussetts: The Harvard Common Press, 1971, 109.
6 Dawkins, Richard. "What Was all the Fuss About?" *Nature*, Vol. 316 (August 22, 1985): 683.
7 Gould, Stephen Jay. *Ever Since Darwin*. Burnett Books, 1978, 161–162.
8 Asimov, Isaac. "14 Million Tons of Dust Per Year," *Science Digest*. (January 1959): 36, cited in Brown, Walt. *In the Beginning: Compelling Evidence for Creation and the Flood*, 7th ed. Phoenix, Arizona: Center for Scientific Creation, 2001, 80.
9 Smith, Elske & Kenneth Jacobs. *Introductory Astronomy and Astrophysics*. Philadelphia, London, Ontario: W. B. Saunders Co., 1973, 101. (Note: one Astronomical Unit (AU) is the distance from the earth to the sun, or about 93,000,000 miles.)
10 Severny, A. B., Kotov, V. A., and Tsap, T. T. "Observations of solar pulsations." *Nature*, Vol. 259, 89, cited in Davies, Keith. "Evidence for a Young Sun." *Impact* #276, El Cajon, California: Institute for Creation Research (June 1, 1996).
11 Ibid.
12 Kazmann, Raphael. "It's about time: 4.5 billion years." *Geotimes* (September 1978): 18, cited in Lubenow, Marvin. *Bones of Contention*. Grand Rapids, Michigan: Baker Book House, 1992, 205.
13 Ibid.
14 Theoretically, as an object that is producing light waves moves away from an observer, the length of the wave is lengthened (in visible light, red light has the longest wavelength, so waves that are lengthened are "red-shifted"). If an object were moving toward an observer, the waves would be shortened or "blue-shifted." This phenomenon, called the Doppler effect, is readily apparent with a source that is producing sound waves—as, for instance, a car moves toward an observer, the wavelengths of the sound will be compacted or shortened, and as the object passes by and moves away, the wavelengths will be stretched or elongated and the observer will hear a distinct change in the pitch of the sound (imagine standing by a racetrack as cars come toward you and then move away).
15 Serway, Raymond A., & Jerry S. Faughn. *Holt Physics*. Austin, Texas: Holt, Rinehart & Winston, 1999, 505.
16 Ibid.
17 Gentry, Robert V. *Creation's Tiny Mystery*, 3rd ed. Knoxville, Tennessee: Earth Science Associates, 1992.
18 Watson, D. M. S. "Adaptation." *Nature* (1929): 233.

Chapter 4

THE BIBLICAL ACCOUNT OF CREATION

One of the concepts that I attempted to ascertain in the previous chapter is that "science" has not established, and cannot establish, the age of the earth at billions of years (especially in light of the fact that it is, in reality, only thousands of years old!). For the most part, it is absolutely impossible for "science" to confirm the age of the earth at all, because the dating process depends on assumptions that are entirely untestable—and, in fact, false. Indeed, the only rocks of known ages (generally volcanic rocks formed within the last two hundred years) that have been "dated" by various radioactive means have yielded highly discordant ages—often in the millions and even billions of years. When that happens, the rock is called an "anomaly" and the particular rock is simply dismissed, yet no one seems to question the dating method itself. Occasionally, scientists will admit that their conclusions about the past are highly speculative, but most often they will not. As Geoffrey Burbidge points out in reference to the big bang theory: "Big Bang cosmology is probably as widely believed as has been any theory of the universe in the history of Western civilization. It rests, however, on many *untested*, and in some cases *untestable*, assumptions. Indeed, big bang cosmology has become a bandwagon of thought that reflects faith as much as objective truth... This situation is particularly worrisome because there are good reasons to think the big bang model is seriously flawed [italics added]."[1] In fact, if you read the conflicting and varying accounts of the different big bang theories, it brings to mind something Mark Twain once wrote: "The researches of many commentators have already thrown much darkness on this subject, and it is probable that, if they continue, we shall soon know nothing at all about it."[2]

Dinosaurs on the Ark

Carved into the granite face of the cliffs at Mt. Rushmore are the faces of four past U.S. presidents. If four geologists, completely ignorant of the history of the rock carvings, were to happen independently upon the faces, one could almost guarantee that they would arrive at four widely varying conclusions as to the age of the carvings and the disposition of the carvers, if all they had to base their calculations on were the rocks themselves. It should be quickly evident, therefore, that the best way to reach conclusions about the carvings would be to have access to historical accounts from the time of the carvings. And that is precisely the situation that we find ourselves in when it comes to the history of the earth and its creation. Fortunately, we were not left in "much darkness" by the Creator, and we have access to historical accounts of both the creation of the world and tidbits of its early history. In fact, there is no such thing as "prehistory" or "prehistorical man" (and there was no such word as "prehistorical" in English dictionaries prior to the nineteenth century). Consequently, it is essential that we gain a clear understanding of the biblical account of creation.

The very first book of the Bible is entitled "Genesis," and in English we have come to attribute the meaning "beginning" to the word. However, that is not its root or its meaning. The Hebrew title of the book is transliterated *bereshith* (בראשית), which does mean "in (the) beginning." Often, in the Hebrew language, the title of a book or document is taken from its opening phrase. In fact, this is a very significant opening because virtually every other culture's creation myths begin with the earth, sun, and matter already existing. But the English title "Genesis" has an even more interesting root. It comes from the Septuagint version of the Old Testament (the translation from Hebrew into Greek). The Greek title is *geneseos* (a root of our English word "generations"), which comes from the Hebrew TOLeDOT. This intriguing term occurs eleven times in the early chapters of Genesis. In the past it has been translated variously as "these are the generations of" or "these are the descendants of." In many places, however, there are no "generations" associated with its use. In fact, the first time it occurs in Genesis 2:4 there

The Biblical Account of Creation

is no name associated with it, and in fact, no person has yet been named in all of creation. Consequently, the NIV translates this occurrence as "this is the account of": *"This is the account of* the heavens and the earth when they were first created." There are no generations either preceding or immediately following this verse and this is the key to discerning its meaning and significance.

Briefly, the eleven locations are as follows: 2:4 (the account of the heavens and earth), 5:1 (the account of Adam), 6:9 (the account of Noah), 10:1 (the account of Shem, Ham and Japeth), 11:10 (the account of Shem), 11:27 (the account of Terah), 25:12 (the account of Abraham's son, Ishmael), 25:19 (the account of Abraham's son, Isaac), 36:1 and 36:9 (the account of Esau), 37:2 (the account of Jacob). In several instances, this phrase is followed by genealogies and the King James translators appropriately translated the term as "these are the generations of." However, to remain consistent, that required translating Genesis 2:4 as "These are the generations of the heavens and earth when they were created, ..." But what could it mean that heaven and earth have "generations," especially as creation had just taken place? Or, even harder to understand, how could heaven and earth have *genealogies* on the very day of their creation? What could the genealogies of heaven and earth imply? Since Genesis 5:1, which the KJV translators rendered, "these are the generations of Adam," was followed by Adam's genealogies, the translators apparently took their cue from this reference and just stayed consistent. They then attached each such verse to the narrative that followed.

On the other hand, the NIV translators translated *geneseos* as "this is the account of." It seems that perhaps the NIV translators may have wrestled with the placement of each of these phrases. Unlike the KJV, the NIV does not attach each phrase to the following section, but neither do they attach it to the preceding section. Instead, the NIV separates each occurrence of *geneseos* as a stand-alone verse, leaving it to the reader to decide. Nonetheless, it is clear that in each instance it was seen as a transition phrase. Yet it seems to me that Genesis 2:4 makes much more sense if it is attached as a summary to the preceding section. The narrative following 2:4 is not the story of the creation of the heavens and the earth, even though it begins with a brief summary; it is, instead, the story of Adam and Eve. In fact, attaching verse 5:1, which should be best translated "this is the written account of Adam," to the narrative preceding, it now makes perfect sense, because it becomes the signature of Adam at the end of his own narrative—a story that only he could have told.[3]

Unfortunately, adding to the confusion surrounding the meaning of this simple phrase, the NIV translates the verse "this is the written account of Adam's *line*." The apparent reason for such a rendering is that the KJV tradition of translating *geneseos* as "generations," as well as attaching each phrase to the section following, was too much tradition to forgo. Furthermore, the generations following Genesis 5:1 extend well beyond Adam's death, so if this phrase is attached to what follows, its authorship cannot be relegated to Adam himself. Consequently, the NIV translators apparently felt constrained to add the word "line" after Adam's name. The fact is, however, there is no word for "line" or "descendants" in the Hebrew text of Genesis 5:1, and the straightforward rendering would be simply "this is the written account of Adam," or "this is the book that Adam wrote." From verse 2:5 through 5:1 is the story that only Adam and Eve could have told as eyewitnesses and participants. I would contend that this rendering is the best translation of this phrase in Genesis 5:1: it is the signature of Adam at the end of the account that he wrote and passed on.

The only occurrence of this term, for which there is no name associated, is its first occurrence in Genesis 2:4. I submit that there is no name associated with this occurrence precisely because the Author was the Nameless One Himself. The information preceding Genesis 2:4 could only have been provided by God Himself (and this is a very significant point). Therefore, each of these ten occurrences (Note: there are actually eleven instances

The Biblical Account of Creation

where *geneseos* is used, but two of them are used with Esau: 36:1 & 36:9) of the Greek *geneseos*, or the Hebrew *TOLeDOT*, would best be seen as the author's signature at the end of his own story. Each of these places represents a clear transition in the narrative, and it would seem highly probable that each of these represents the written account owned and even authored by the person at the end of each section. That means the version of events from Genesis 1:1 through 2:4 was authored directly by God and given to Adam. In which case, dismissing or downplaying its significance by calling it "myth" or "poetry" is not only foolish, it is arrogant. We need to give serious consideration to its accuracy and purpose, authored as it was by the Creator Himself:[4] "This is the account of the heavens and the earth *when they were created* [italics added]." It was not only firsthand, but the written expression of it was timely! If there are differences or disagreements between the biblical creation account and a naturalistic explanation (there are numerous and significant ones), there can simply be no compelling reason for accepting pronouncements of finite men who were not there, over the revelation of the Infinite One who was there!

A very significant discovery that supports this understanding was published in 1936 by P. J. Wiseman (*New Discoveries in Babylonia*). He discovered an ancient Babylonian literary device called a *colophon*. This is "a scribal device placed at the conclusion of a literary work written on a clay tablet giving ... the title or description of the [preceding] narrative, the date or the occasion of the writing, and the name of the owner or writer of the tablet."[5] Most striking about the colophon and its parallels to the divisions of Genesis is that the name in the colophon often signified the owner or the writer of the tablet. And often they were one and the same. The colophon also often included the date or the occasion of the writing (cf. 2:4, "when they were created," or 37:12, "when he lived in Canaan"). And remember that Hebrew is also a language that is written in what (to us) might be considered backward fashion, right to left, starting at the back of the book, so the colophon would actually end up being on the front page (or tablet) to most easily identify its ownership or authorship. In every case, in Genesis, the person named could have written the contents of the preceding section from their own experience or from sources available to them, but not necessarily the following section. As Lubenow concludes, "The implications of this evidence for the origin of Genesis are staggering. Rather than Genesis having a late date, as is universally taught in non-evangelical circles [and nowadays, in many evangelical circles], it implies that Genesis 1-11 is a transcript of the oldest series of written records in human history."[6] It means that "the actual authors of Genesis were [God], Adam, Noah, the sons of Noah, ... Terah, Ishmael, Isaac, Esau, Jacob, and Joseph ... and that Moses, utilizing these records, was the redactor or editor of Genesis rather than its author."[7]

In support of this conclusion, I think it is telling to look at the account of Noah and the flood. In Genesis 6:9, the third occurrence of *geneseos* is found: "This is the account of Noah." Though the preceding narrative is not the account of the flood, it is the brief history of the earth from Adam through Noah that Noah could reasonably have been responsible for compiling or keeping. The verse preceding the use of the apparent colophon states very simply, "But Noah found favor in God's eyes." It strikes me that this is not only true, but it is consistent with Noah's character and the way he might speak of himself—he was humbled that he found favor with God. Yet at the beginning of the next narrative in the second half of verse 9, Noah is reintroduced as "a righteous man, blameless among the people of his time, and [a man who] walked with God." Why the new introduction only moments after the last? And why the difference? It makes perfect sense that, following the colophon that indicates the signature of Noah, the narrative of Noah's sons, Shem, Ham, and Japheth begins. Their introduction of Noah, though still true, is less humble, precisely because it is no longer a first-person narrative. They are writing—and, perhaps, bragging!—of their father, Noah, who was a righteous and blameless man. Their narrative

The Biblical Account of Creation

ends in Genesis 10:1 with "this is the account of Shem, Ham, and Japheth, Noah's sons, who themselves had sons after the flood." Since Genesis 9:29 tells of Noah's death, it should be clear that the reference to "the account of Noah" in 6:9 could not imply that he wrote the narrative that followed and told of his own demise. And of course, Noah would undoubtedly have been a source of information that his sons used in the recording of the narrative of the flood.

Though II Timothy 3:16 states that "all Scripture is God-breathed," modern scholarship has seemingly been infected by naturalistic or evolutionary ideas when discussing the authorship and accuracy of Genesis by assuming that writing "evolved" over centuries or even millennia. The older the text, the less trustworthy it is deemed because it was undoubtedly maintained by oral tradition for lengthy periods before being committed to written form. And yet, scripturally, the opposite may very well be true. The sources in Genesis were not only "God-breathed," but they were firsthand accounts by men far closer to the perfection of created form than now. The Bible is not only true history, it is His Story, breathed through His chosen vessels, and there is evidence that even the manner of telling is unique to all other written accounts.

A number of years ago in Czechoslovakia, a rabbi, H. M. D. Weissmandel, noticed that if he took the first occurrence of "T" in Genesis (Hebrew: ת) and then skipped fifty letters, took the next letter, skipped fifty, and so on, it spelled out "Torah," the Hebrew word for "the book of the law." He thought it was an interesting phenomenon, but assumed it was probably a fluke, until he discovered that the same thing occurred in Exodus, Numbers, and Deuteronomy. Isaac Newton, brilliant physicist and mathematician, spent the last several years of his life looking in the Scriptures for evidence of a hidden message that would give credence to its divine authorship, though he was unable to find what he sought. It took the advent of modern supercomputers to find such evidence.

Several years ago, a Jewish mathematician decided to look for more examples of information encoded within the text of the Old Testament. He took the Hebrew text, stripped of all its vowels and word divisions (as it was originally) and began a computer search. He took the names of thirty-two great wise men (from biblical to modern times) along with the dates they were born and died and had the computer do random sequential letter skips. Much to his surprise, he found all the names encoded within the text. He then tried to duplicate the results using the Hebrew text of *War and Peace*, as well as two other original Hebrew texts. He found nothing. He then took the names and dates and jumbled them up in ten million different combinations, only one of which was correct. He found only the correct version "encoded" within the biblical text. Harold Gans, a senior code-breaker with the National Security Agency who also spoke and read Hebrew, learned of Rips's work and thought it was "off-the-wall ridiculous," so he set out to debunk it. He thereafter repeated the experiment by adding another thirty-four names to the original list, for a total of sixty-six names. He also decided that, if the "code" was legitimate, he might be able to add the cities in which these men were born and died, and find these too. He found them all. The results, he said, "sent a chill up my spine."[8] He went on to conclude "that these results provide corroboration of the results reported by Witztum, Rips, and Rosenberg ... At first I was 100% skeptical, ... I thought this was all just silly. I set out to disprove the code, and ended up proving it."[9] The work has been peer-reviewed by skeptical, disbelieving mathematicians and they have found no flaws. Their conclusion: "The Bible is encoded with information about the past and about the future in a way that was mathematically beyond random chance, and *found in no other text* [italics added]."[10] Yale mathematics professor Piatetski-Shapiro has studied the results and come to the conclusion that "what we're talking about here is some intelligence that stands outside. I think it is the only answer—that God exists."[11]

No accidental or coincidental explanation can suffice because, to date, no one has been able to replicate the phenomenon in any text other than the Hebrew Old Testament, and many now believe it could potentially be more complex than we have yet been able to discover. Mathematicians and scientists who have actually studied it and duplicated the test results are agreed that all the computers currently existing in the world working together could not have encoded the Bible in the way it apparently was over three thousand years ago. Eli Rips stated succinctly, "I can't even imagine how it would be done, how anyone could have done it. It is a mind beyond our imagination."[12]

Understanding that the creation account in Genesis 1 is not only unique among the world's creation accounts, but that it was written by the Creator Himself, should cause us to take seriously its content and meaning, whether it is encoded or not. And even a brief glance demonstrates that its version cannot be reconciled with any "scientific" attempt to explain the beginning of the earth and universe.

On the first day, the "heavens and earth" were created, and that was followed by the creation of light and its separation from darkness. The light was called "day" and the evening called "night," and "there was evening, and there was morning—the first day." The phrase "there was evening, and there was morning" is attached here, as it is every other day, precisely to show that this day was the same as every other day, and it is a phrase that in Hebrew indicates a normal, twenty-four-hour day. It should already be clear that no naturalistic or "scientific" explanation of the beginning of the universe begins with the formation of the earth—before there were stars—or with light and darkness before there was even a sun. But God does not require a sun to form light, nor does the existing sun prevent him from producing darkness if it suits him. God himself is light and he requires no source to produce it. John proclaims, "This is the message we have heard from him and declare to you: God is light; in him there is no darkness at all [1 John 1:5]." Recall that one of the ten plagues in Egypt was simply a plague of darkness that endured for three days. In fact, it was called a darkness "that can be felt [Gen 10:21]." "No one could see anyone else or leave his place for three days. Yet all the Israelites had light in the places where they lived [Gen 10:23]." This was not some kind of eclipse or "natural" manipulation; it was the supernatural act of the Creator separating light (over Goshen) from darkness (over Egypt).

On the second day, God created an "expanse" between the waters that separated the waters above from the waters below. The "expanse" was called "sky," so it would appear that there was some kind of water—or water-vapor—canopy that surrounded the earth at its creation. Certainly it is clear from many aspects of the fossil record that the earth's atmosphere before the flood was very different from the one that now exists. Plant life, as well as animal life, grew much larger and in much more abundance. Other studies have also indicated that quite likely the atmospheric pressure was greater and there was a higher percentage of oxygen in the preflood atmosphere. No naturalistic scientific guesses about the earth's past have postulated such a thing.

On the third day, God created dry land and seas (note: the separation of the dry land and seas must have been different from the previous separation that had water above the "expanse"), then vegetation—plants and trees. Again, no "scientific" or naturalistic theory has postulated the evolution of plants and trees before the existence of the sun. At the end of both the second and the third days, God pronounced everything to be good. Yet all attempts to harmonize vast ages of time with the biblical creation account (cf. following explanations of the "great gap theory," and the "day-age" theory) require that Satan had already fallen, and millions or billions of years of death, destruction, chaos and disease were already part of the earth at this point in creation. It is impossible to conceive of the perfect Creator looking down at such a mess and calling it "good." Especially when, after the sixth day and the creation of man, he called it "very good."

On the fourth day, God created lights in the sky to serve as signs to mark the seasons, days, and years

The Biblical Account of Creation

(note: light was already present). The sun, moon, and stars were made on the fourth day, and since "stars" and "angels" are often used interchangeably in Scripture[13], I would conjecture that the angels were created on the fourth day as well. Satan is a created being and he is not preexistent. He therefore could not have fallen millions of years before the creation of the earth or of Adam and Eve.

On the fifth day, God created all sea/water creatures as well as flying creatures. This means that according to Genesis, birds existed before air-breathing, land animals, which would necessarily include the dinosaurs. Once again, all evolutionary or long-age "scientific" accounts have dinosaurs existing for millions of years before birds came into being (and one of the more popular current myths has dinosaurs evolving into birds). Again, the text says clearly that God saw that all that he had done, and all that existed, and it was "good," but every long-age prognosticator (whether atheistic evolutionist or "progressive creationist") must insist that millions of years of death, disease, killing, and chaos had already taken place by this time and therefore must be part of God's idea of "good."

Finally, on the sixth day, God created all air-breathing, land creatures: "livestock, creatures that move along the ground, and wild animals [Genesis 1:24]," and then made man as male and female. And, after the sixth day of creation, "God saw all that he had made, and it was *very good* [Gen 1:31]." Once again, if any type of long-age creation took place, then millions or billions of years of death, chaos, decay, and mass destruction is not only part of God's idea of "good," but it is "very good." Both progressive and theistic creation theories completely nullify the effects of the curse being a result of man's sin. The earth had already been cursed by death, destruction, and decay for untold millennia. Yet following Adam and Eve's sin, Genesis 3:17-18 says, "Cursed is the ground because of you... It will produce thorns and thistles for you." Not even the painful thorns of the beautiful rose were part of the "good" of God's creation, but one of the effects of the curse of sin. Romans 5:12-14 states, "Just as sin entered the world through one man, and death through sin, and in this way death came to all men ... death reigned from the time of Adam to the time of Moses, even over those who did not sin by breaking a command ..." As Dr. F. Godet puts it, "This reign of death which prevails over all that is born cannot be the normal state of a world created by God. Nature suffers from a curse which it cannot have brought upon itself, as it is not morally free."[14] Is it not ironic that the thorns that were produced as the effect of man's sin were the very things that were wrapped around the brow of the One who came to pay the penalty of our sins? Ironic it may be, but accidental it was not!

One of the two primary theories that attempts to harmonize vast eras of time with the biblical creation account has been referred to as the "day-age" theory. Though it has a few variations, it basically postulates that each "day" of creation was actually a long period of time. Yet as a method of harmonizing Scripture with "science" such conjecture falls woefully short. If the days of creation represent long ages of time, then the sequence of events recorded in Genesis is completely impossible. By this account, plants (created on day three) had to have existed for millions of years before there was a sun to sustain them (day four). "Progressive creationist," Hugh Ross attempts a very convoluted harmonization that has creative "periods" overlapping each other so that the main creation of stars took place during the fourth time period, but he then claims that many stars were previously created.[15] As Ross states, "To insist that the creation events of Genesis must be specifically limited to the creation days in which they are mentioned is to read too much into the text. It can more reasonably be stated that each event or life-form was primarily introduced on the creation day indicated and that each creation day was pre-eminently one of the introduction of the event or life-form(s) described."[16] This twisted rendering has absolutely nothing whatsoever to do with the actual text of Genesis, and it completely ignores the fact that the earth was created on the very first day of creation. Were there several

The Biblical Account of Creation

earth creations that "overlap" with star creations? The text itself gives absolutely no hint that it would somehow be more "reasonable" to assume a creation order and process other than what is indicated. It is quite clear from Genesis that plants were created on day three, and the sun not until day four. There is no conceivable way to postulate the creation of the sun as an "overlapping" event that "primarily" took place on the fourth creation "day" but that also preceded the creation of plants that "primarily" took place on the third "day" and then pretend that this is somehow a more "reasonable" interpretation of the events recorded. Every theory that includes long ages of time (whether atheistic evolution or progressive creation) requires that stars existed for billions of years before the earth and certainly before plants! But that is entirely at odds with the clear rendering of the Genesis account of creation. It also completely ignores the existence and meaning of the phrase that is attached at the end of each day's creation: "and there was evening, and there was morning." And bear in mind that no one ever arrived at such an incredibly convoluted interpretation of Scripture until the advent of uniformitarian geology (which was an intentional effort to eliminate any interpretation of geological data that gave evidence of a worldwide flood) and evolution (which was an intentional effort to eliminate the evidence of and need for a Creator).

If progressive creation—or anything like it—is true, then these processes were going on for millions upon millions of years before Adam even arrived on the scene. How is it then that the earth and its inhabitants were "cursed" as a result of the sin of Adam in a way that was different from what they were already experiencing? Though some commentators have tried to argue that the curse of death applied only to man and not to animals, it is textually clear that neither men nor animals were created to kill and eat other animals (see chapter "The Scriptural Advent of Animal Carnivorousness")—the world was created as a perfect environment. Furthermore, Ross and others who hold to a vastly old universe must attribute the fossil record to numerous catastrophic events over billions of years of time before the creation of Adam and man's subsequent fall. Yet contained within the fossil record is much evidence of human existence (see chapter "Archaeological Anomalies"). Thorns and thistles are also found entombed within the rocks of the fossil record, and yet these are scripturally attributed to the effects of the Fall (see chapter "Dating a Dinosaur"). If the fossil record is accepted as evidence of vast eras of time, rather than the by product of the flood of Noah, then human beings either existed millions of years before Adam—and were dying before Adam sinned—or Adam existed millions of years ago (which not even Ross dares to postulate). Incredibly, Ross attempts to overcome this dilemma by postulating that there were "human-like" creatures that were able to communicate, make instruments and tools, live in communities, and even bury their dead, but had no souls, so they were not really "human." Finding someone who would believe such a bizarre rationalization seems remarkable, but not nearly as incredible as finding someone who claims that Scripture can be used to support such a concoction.

Unfortunately for Ross and others of his ilk, in the historical context of Genesis 1, the meaning of the word "day" is neither unclear nor ambiguous. The Hebrew word for "day" used in Genesis 1 is transliterated *yom*. According to Morris, it can have basically three meanings: a solar day (twenty-four hours), daylight (sunrise to sunset), or an indefinite period of time. It occurs well over two thousand times in the Old Testament, and it almost always means a literal day. But when it is used in the plural form, *yamim*, it always refers to literal days (over eight hundred times). Furthermore, when it is modified by a number in historical narrative (359 times outside of Genesis 1), there is no known occurrence when it can mean anything other than twenty-four-hour days. The addition of the modifying phrase "evening and/or morning" (thirty-eight times outside of Genesis) is tantamount to emphasizing that these are literal days, and there is no other possible, let alone plausible, intepretation.[17] In order to translate

The Biblical Account of Creation

"day" as something other than a literal day in Genesis 1, every known exegetical guideline would have to be broken or ignored. There is absolutely no textual reason for doing this. Neither is there any hermeneutical principle that would even allow for it.

As additional textual support for the meaning of the word "day" in the context of Genesis, Exodus 20:11 states that "For in *six days* the Lord made the heavens and the earth, the sea, and all that is in them ..." There is no place in any known Hebrew literature where this could mean six indefinite or overlapping time periods. If that is what was intended, there are many ways it could have been said differently in the Hebrew language to make this clear. For instance, the Hebrew word *olam* means a long or indefinite period of time. If God wanted us to be clear that creation was accomplished in six literal twenty-four-hour periods, there is no clearer way to say it in the Hebrew language. In fact, if Exodus 20:11 does not mean that creation occurred in six literal days, then there would apparently be no way in the Hebrew language to say such a thing. What compelling textual reason could there possibly be to interpret it so completely contrary to any known rule of hermeneutics or exegesis? Furthermore, recall that Exodus 20:11 is in the midst of the Ten Commandments that were written in stone by the hand of God Himself. Exodus 31:18 states that "when the Lord finished speaking to Moses on Mount Sinai, he gave him the two tablets of the Testimony, the tablets of stone *inscribed by the finger of God* [italics added]."[18] When one considers that the Creator of the universe wrote in stone by his own hand that everything was created in six days—knowing that for thousands of years this could not be understood as anything other than literal days—it should be taken with sober-minded seriousness.

The second rationale for inserting vast eras of time into Scripture is called the "great gap" theory. This theory basically states that the earth and the dinosaurs were around for millions of years; then Satan fell and the earth was destroyed and recreated. All this took place between verse 1 and 2 of Genesis 1. This interpretation finds its textual support from translating the verb in verse two as "became" instead of "was" (i.e., "the earth *became* formless and empty"). Such a translation is not suggested by the context, nor is it translated that way in any of the current English versions. In fact, such a translation was never suggested until the modern attempts to find some place to fit vast periods of time into a Scripture that has no other place for them. In any case, such an attempt at harmonizing the biblical account of creation with vast eras of time fails for all the same reasons previously addressed.

Arthur C. Custance attempted to offer a defense of the gap theory by stating, "If a vast antiquity far beyond the 4000 BC traditional date is demanded [by whom!], there are other ways in which a great antiquity for the world prior to the creation of man can be allowed for. For example, the days of Genesis might be viewed as days on which revelation was given to Moses; or they might be taken to mean ages; or we might introduce a hiatus between Genesis 1:1 and 1:2 and so on."[19] (Note: Is this not tantamount to saying we can make Scripture say whatever we need it to, rather than determining what Scripture actually does say?) Nevertheless, he went on to say, "I do not think that the biblical account can ever be made to accommodate the antiquity that is still being demanded. Personally, I am convinced that the arguments for this vast antiquity will in due course be modified by fresh evidence and the Bible validated as it always has been."[20]

Dr. Pattle P. T. Pun admits that "It is apparent that the most straightforward understanding of the Genesis record, without regard to all the hermeneutical considerations suggested by science, is that God created heaven and earth in six solar days, that man was created in the sixth day, that death and chaos entered the world after the Fall of Adam and Eve, that all of the fossils were the result of the catastrophic universal deluge which spared only Noah's family and the animals therewith."[21] Incredibly, Dr. Pun recognizes that this is true, but nevertheless accepts the prognostications of secular science over the revealed truth of Scripture. Furthermore, as Dr. Davis

The Biblical Account of Creation

Young points out, "It cannot be denied, in spite of frequent interpretations of Genesis 1 that departed from the rigidly literal, that the almost universal view of the Christian world until the eighteenth century was that *the Earth was only a few thousand years old*. Not until the development of modern scientific investigation of the Earth itself would this view be called into question within the church [italics added]."[22] Again, even though Young himself recognizes this historical persuasion, he subscribes to an old-earth position that removes historical and factual content from Genesis. The reasons for doing this are not textual, hermeneutical, or even historical, but "scientific."

Though the exact date of creation may not be attainable (and yet, it may) through analysis of the biblical genealogies and time periods in Scripture, most scholars who have utilized the Scriptures have arrived at very consistent calculations. According to John Calvin, "But little more than five thousand years have passed since the creation of the universe ..."[23] Martin Luther wrote that "we know from Moses that the world was not in existence before 6,000 years ago."[24] Brilliant physicist and mathematician Isaac Newton wrote an indictment of ancient Egyptians because they allowed for events before 5000 BC, which Newton knew Scripture did not allow for.[25] Numerous scholars and historical sources have previously come to the same conclusion for the biblical date of creation: the Jewish calendar states 3760 BC, Johannes Kepler calculated 3993 BC, Melanchthon 3964 BC, Martin Luther 3961 BC, Lightfoot 3960 BC, and Playfair 4008 BC.[26]

In actual fact, the idea of an earth that is billions of years old has never been gleaned from a straightforward study of either the context or language of Scripture. And though there were pagan myths and philosophies that considered the earth vastly old, and there were even previous attempts by liberal theologians to extend the time allowed for in Scripture, most scholarship continued to contend for the literal historicity of Genesis until the mainstream acceptance of the writings of Charles Lyell and Charles Darwin (two people who, by their own written admissions, were attempting to eliminate the Bible as a source for history or Truth). Archbishop James Ussher was a distinguished theologian and scholar who had even been allowed to address Parliament and the king. According to The Evangelical Dictionary of Theology, "He was much sought after by contemporaries for his knowledge and beauty of character, and his personal impact was probably even greater than his scholarly legacy."[27] He used the biblical chronologies to place the creation event accurate to the day. (Note: Whether or not the biblical chronologies were intended to be accurate to the day, it should be astonishing and revelatory that anyone could even attempt to do so, since no one has ever dared to suggest that the Scriptures reveal a "creation" occurring 4.6 billion years ago!) Ussher's detailed and thorough analysis of the Hebrew text led him to conclude that creation occurred in 4004 BC. No modern scholarship has found any significant flaws in his work, nor any way to improve on the accuracy of his conclusions.

To illustrate that the liberalizing in biblical chronology interpretations occurred largely post-Lyell and Darwin, "one concise and readily available source of nineteenth-century information is Robert Young's concordance, and in the popular twenty-second edition, under 'creation,' will be found a list of thirty-seven computations of the date of creation ... Of these thirty-seven, thirty are based on the Bible and seven are derived from other sources—Abyssinian, Arab, Babylonian, Chinese, Egyptian, Indian, and Persian. Not one of these ancient records puts the date of creation earlier than 7000 B.C. In all the hundreds of thousands of years over which hominid man is alleged to have evolved, it is surely more than coincidental that ancient civilizations, which were by no means ignorant of timekeeping by astronomical methods, should all begin their historical record at this arbitrary date. In addition, all myths and legends, however bizarre, speak of instant creation just a few thousand years earlier."[28] It should also be instructive to note that no written human records exist of any earlier time period.

The Biblical Account of Creation

Charles Berlitz, who categorically believes in the vast eras of time postulated by evolutionary "science," nevertheless admits that the date of creation given by Ussher "oddly enough, corresponds vaguely to a written history starting point in the Egypt-Sumeria area."[29]

In fact, premier Hebrew scholar James Barr states, "Probably, so far as I know, there is no professor of Hebrew or Old Testament at any world-class university who does not believe that the writer(s) of Genesis 1-11 intended to convey to their readers the ideas that (a) creation took place in a series of six days which were the same as the days of 24 hours we now experience (b) the figures contained in the Genesis genealogies provided by simple addition a chronology from the beginning of the world up to the later stages in the biblical story (c) Noah's flood was understood to be world-wide and extinguish all human and animal life except for those in the ark. Or, to put it negatively, the apologetic arguments which suppose the 'days' of creation to be long eras of time, the figures of years not to be chronological, and the flood to be a merely local Mesopotamian flood are not taken seriously by any such professors, as far as I know."[30] If it is clear what the author(s) of Genesis intended to convey, and the words were "God-breathed," then to vary their intended meaning now by assigning them new and unintended meanings is tantamount to rewriting Scripture.

It is absolutely essential in this age of "reason" and compromise that we return to a biblical foundation and sound hermeneutical principles. Evangelical author Dr. Gleason Archer states: "From a superficial reading of Genesis 1, the impression would seem to be that the entire creative process took place in six twenty-four-hour days. *If this was the true intent of the Hebrew author* ... this seems to run counter to modern scientific research which indicates that planet earth was created several billion years ago [italics added]."[31] Someone needs to ask, "If this was not the true intent of the Hebrew author, then why did he leave us with that clear impression?" Was he merely ignorant of how to say what he meant (even under the inspiration of the Holy Spirit), or was he truly ignorant of the facts as well? As Barr has previously pointed out, no serious Hebrew scholar denies what the intent of the author was. Some modern Christians (such as Ross) have even gone so far as to suggest that, while this may have been the "intent of the author," the author was ignorant of the real meanings of his own words! Does this apply even when the author was the Creator Himself? How does this principle work when we apply it to the New Testament? How about using it in the Gospels? Taking Archer's hermeneutic principle and applying it to the New Testament, one could reasonably conclude: "From a superficial reading of the Gospels, the impression would seem to be that the Crucifixion literally took place, that Jesus was really dead and the Resurrection actually took place on the third real day. If this were the true intent of the Greek author, this seems to run counter to modern scientific research that indicates that dead bodies do not come back to life and actually tend to stink after three real days in a tomb." We compromise and belittle the foundational truths of the Word of God when we try to reinterpret its clear and historical meanings in light of the so-called scientific pronouncements of men.

As I see it, to interpret Genesis 1 in any way other than the obvious requires three things: (1) you must be willing to put more faith in the so-called science of man (many of whom historically intended by their "science" to falsify Scripture) than in the Revelation of God, (2) you must believe that the authors of Genesis were ignorant of the real meanings of their own writings, and (3) God's revelation in Scripture was not only misinterpreted, it was in fact unable to be interpreted correctly for thousands of years, until the advent of uniformitarian (or naturalism) and evolution theories (two theories whose spiritual roots are demonstrably antibiblical).

It should also be noted that the genealogies of Genesis (chapters 5 and 11) are, to my knowledge, unique in all of ancient literature. To sustain the "day-age" theory many have tried to argue that the genealogies are incomplete or that the word translated "son" might mean grandson or even descendant. (Note: even if several generations were missing between every set of names mentioned, there is no way to add even millions of years to the scriptural account, let alone billions.) However, even if it could somehow be demonstrated that there was any validity to such claims (it cannot), it is an entirely moot point, because the age of every single patriarch is given at the time when the descendant through whom the lineage is being traced is born (see illustration below).

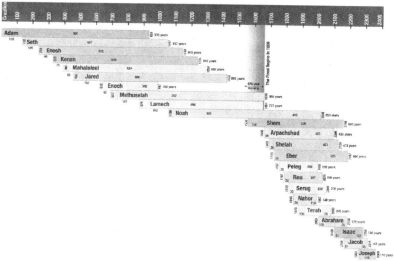

Patriarchal timeline. [Fig. 183, *In the Beginning: Compelling Evidence for Creation and the Flood*, 8th ed., 2008, by Walt Brown, p. 380, Bradley W. Anderson and Walt Brown.]

Not one is missing. Any honest attempt to add millions of years to the biblical narrative—where they clearly do not exist—must eventually deny the accuracy or authority of Scripture. If the numbers in these genealogies are simply unimportant or fictional numbers, why were they even included at all? Some prognosticators have tried to reinterpret the reported ages of the preflood patriarchs because they believe it is "scientifically" unreasonable to think that humans actually lived for over nine hundred years. Consequently, it has been postulated that the preflood word for "year" may have actually meant "month." There is clearly no textual precedent for such a thing and it quickly leads to impossible contradictions. First of all, Mahalalel (5:15) is reported to have had his son at sixty-five. If the word "year" here means "month," then Mahalalel was five and a half years old when he had his first son. Second, in the years following the flood, the ages of Noah's descendants quickly diminish with each passing generation—it follows a mathematical exponential decay curve very nicely. It quickly becomes apparent that there is no reasonable point at which to insert the change in meaning from "month" to "year." Even Abraham, the eighth generation following the flood, lived to be 175 years old—quite "scientifically" impossible by today's standards, but even more ludicrous if one changes the meaning to 175 "months" and concludes that Abraham died before his fifteenth birthday!

There is no intrinsic reason for reinterpreting the clear and concise meaning of the creation account in Scripture—not the length of the days, not the order of the creation events, and certainly not the intact and precise genealogies that extend from Adam all the way to Abraham and beyond. Compromises with the world of secular science, such as the great gap and day-age theories, are remarkably like the theory of evolution: they are both capable of adapting to virtually any change in pronouncements from science, but they have predicted nothing, and they all produce glaring contradictions with different aspects of the clear meanings of Scripture. No one has yet been able to glean from Scripture the actual "old-age" of the earth of 4.6 billion years (or any other previously believed "age"). Indeed, should some scientists decide tomorrow that the actual age of the earth is 6.4 billion years, either theory could quickly adapt, but neither will ever predict such a thing. In fact, the only prediction I have seen (from either progressive creationists or atheistic evolutionists) is that there is no ark on the top of Mt. Ararat. Perhaps that is one prediction that is falsifiable.

Colossians 2:8 says, "See to it that no one takes you captive through hollow and deceptive philosophy, which depends on human tradition and the basic principles of this world rather than on Christ." To attempt to use man's interpretations of nature over and above God's clear revelation is to allow room for deception. As Dr. Charles Ryrie states succinctly, "God Himself is the Source of our knowledge of Him... Only true truth comes from God, for since sin entered the stream of history man has created that which he calls truth but which is not. Furthermore, he has perverted, blunted, diluted, and corrupted that which was originally true truth that did come from God. For us today the only infallible canon for determining true truth is the written Word of God. Nature, though it does reveal some things about God, is limited and can be misread by mankind. The human mind, though often brilliant in what it can achieve, suffers limitations and darkening."[32]

Endnotes

1. Burbidge, Geoffrey. "Why Only One Big Bang?" *Scientific American* (February 1992): 120.
2. Twain, Mark, quoted by Ken Croswell in "The Constant Hubble War." *New Scientist*, Vol. 137 (February 13, 1993): 23, cited by Henry Morris in *That Their Words May Be Used Against Them*. Green Forest, Arkansas: Master Books, 1997, 6.
3. Genesis 5:1 is the only occurrence of this phrase (TOLeDOT) where the Hebrew word for "book" or "written account" occurs, thus demonstrating that this was not just a later "story" attributed to Adam, but the account that he actually wrote and passed down.
4. Modern scholarship has tended to take a dim view of Adam's writing ability, but recall that God created him as perfect man with the ability to communicate immediately and fully with his Creator. Also recall that God wrote the Ten Commandments "in stone with his own finger" and gave them directly to Moses (cf. Gen. 31:18 and 32:16), so there is no reason to postulate that the Creator would not have written this very important firsthand account himself, or that Adam could not have read it.
5. Lubenow, Marvin L. *Bones of Contention: A Creationist Assessment of the Human Fossils*. Grand Rapids, Michigan: Baker Book House, 1992, 217.
6. Ibid., 221.
7. Ibid., 217-218.
8. Drosnin, Michael. *The Bible Code*. New York: Simon & Schuster, 1997, 22-23.
9. Ibid., 23-24.
10. Ibid., 24.
11. Ibid.

12. Ibid., 46.
13. For example, in Isaiah 14:12-13, Lucifer is apparently referred to as the "morning star": "How you have fallen from heaven, O morning star, son of the dawn! You have been cast down to earth, ... You said in your heart, 'I will ascend to heaven; I will raise my throne above the stars (angels) of God'..."
14. Godet, F. *Commentary on St. Paul's Epistle to the Romans*. New York: Funk & Wagnells Publishers, 1883, 314, cited in VanBebber, Mark, & Paul S. Taylor. *Creation and Time: A Report on the Progressive Creationist Book by Hugh Ross*. Gilbert, Arizona: Eden Communications, 1994, 52.
15. Ross, Hugh, Ph.D. *Genesis One: A Scientific Perspective*, rev ed. Sierra Madre, California: Wisemen Productions, 1983, 12.
16. Ibid.
17. Morris, John D., Ph.D. *The Young Earth*. Colorado Springs, Colorado: Master Books, 1994, 29.
18. See also Exodus 32:16. Though Moses destroyed the first set of tablets, God told him to carve out another set of tablets and return to Sinai, where he promised to re-write the commandments.
19. Custance, Arthur C. *Two Men Called Adam*. Brockville, Ontario: Doorway Publications, 1983, 246, cited in Brown, Walt, Ph.D. *In the Beginning: Compelling Evidence for Creation and the Flood*, 7th ed. Phoenix, Arizona: Center for Scientific Creation, 2001, 275.
20. Custance, 249, cited in Brown, 275.
21. Pun, Dr. Pattle P. T. *Journal of the American Scientific Affiliation* (March 1987): 14, cited in Morris, 30.
22. Young, Davis A. *Christianity and the Age of the Earth*. 1982, 25, cited in Morris, 29.
23. Calvin, John. *Calvin's Institutes of the Christian Religion*, Vol. 2. Edited by John T. McNeill. Philadelphia, Pennsylvania: Westminster Press, 1960, 925, cited in Van Bebber and Taylor, 120.
24. Peliken, Jaroslav, ed. "Luther's Works." *Lectures on Genesis Chapters 1-5*, Vol. 1. St. Louis: Concordia Publishing House, 1958, 3, cited in Van Bebber & Taylor, 103.
25. Morris, Henry M. *The Genesis Record: A Scientific & Devotional Commentary on the book of Beginnings*. Grand Rapids: Baker Book House, 1976, 44.
26. Ibid., 45.
27. *The Evangelical Dictionary of Theology*. 113, cited in Van Bebber & Taylor, 102.
28. Taylor, Ian T. *In the Minds of Men*. Toronto: TFE Publications, 1984, 283, cited in Ham, Ken. *The Great Dinosaur Mystery Solved*. Green Forest, Arkansas: Master Books, Inc., 1998, 115.t
29. Berlitz, Charles. *Mysteries From Forgotten Worlds*. New York: Del Publishing Company, Inc., 1972, 7.
30. Barr, James in a letter to David Watson, 1984, cited in Morris, *The Young Earth*, 31.

31 Archer, Gleason L. *A Survey of Old Testament Introduction*. Chicago: Moody Press, 1979, 181.
32 Ryrie, Charles. *Basic Theology*, 25, cited in Van Bebber & Taylor, 29.

Chapter 5

1998 AND 1999 SEARCHES FOR THE ARK

After I'd met Richard Bright in Dogubayazit in 1997, we kept in loose contact throughout the year and ultimately, he invited me to accompany him to Turkey in 1998 for some potentially clandestine attempt to climb the mountain. Efforts had been made to work with "official channels" and a Turkish university professor named Salih Bayraktutan. He claimed to have an inside track to research permits on Ararat. Yet, after I journeyed halfway around the world for a second time chasing promise of "permits," it once again proved to be an elusive hope. The hoped-for permits with Dr. Salih Bayraktutan never materialized. Richard and I made several local contacts and schemed schemes, but in the end it was all to no avail. At one point, we were actually packed and on the way to the mountain, or so we thought. We had previously made contact with a military officer who was willing to provide us with safe passage up the mountain. Once at the top, we were to be given forty-eight hours "on our own" to go where we wanted. If we were arrested on the way down—having been abandoned by our guide forty-eight hours earlier—we were to say that we were guests of the officer. That was our "guarantee." At the contact point we waited in vain. Eventually we received a phone call and were told that it was now "impossible" to go. Supposedly, a military operation had begun and we were informed that it might last for several days. It was all very frustrating to say the least.

With no climbing immediately forthcoming, through some more local contacts we were able to set up a meeting with the Turkish general in Agri who was responsible for overseeing the military in that region of the country. Our meeting was all very pleasant and cordial, and even though little of the conversation was actually translated into English for our benefit, it was clear from the outset that we were not gaining anything in being there.

Consequently, before we left, I asked that my questions be translated to the general (though I suspect he understood my English well enough—few get to the position of general in the Turkish military without having command of the English language). I said that the American military had photographed the ark. I personally knew of at least three trustworthy people who had seen the ark. We knew it was there, and it could not be kept hidden indefinitely. In lieu of the foregoing, I asked, would it not be better for the military of Turkey to be seen by the world as participants in finding one of the greatest archaeological treasures of all time, rather than to be eventually revealed as those who had intentionally obstructed its discovery? He seemed outwardly nonplussed by the question, but he responded simply by asking if he could come to America and climb our mountains. I replied that I would gladly escort him to the top of any mountain of his choice—for free! I pointed out that other countries with archaeological sites (like Egypt and the pyramids) allowed both public access and historical research to take place. "Why," I asked, "is Turkey afraid of the discovery of Noah's Ark?" If he had an answer, he did not proffer it. It was interesting, though, that when we left the meeting, the owner of the Hotel Ararat (who had accompanied us as our liaison) turned to me, shook my hand, and said through the translator, "David, thank you for your very excellent question."

In the end, I wondered what I was to learn from each separate experience. Was God telling us that this was not the way to do things? Or that now was not the time? Or that not enough prayer warfare had been laid as foundation? My one consolation was, and continues to be, that virtually everyone involved in the search over the years has experienced similar frustrations, setbacks, and unanswered questions. I remain more convinced than ever that God has preserved a significant portion of the ark for just such a time as this. He will allow its exposure in "the fullness of time." But, just as the flood was delayed until Noah had completed the ark, we must continue to do our part. And we will do well to remember that we war not against flesh and blood, but against principalities and powers.

In 1999, I was initially asked by John McIntosh to be a part of his small "Omega Team," which was to include Richard Bright and Paul Thomson, as well as John's friend and contact in Turkey who claimed he could get us permission for a climb. However, while awaiting permission, John was asked to be a part of a larger effort, dubbed the Ark Research Project (ARP) and headed by Jim Hall. John soon decided to concentrate most of his efforts with ARP, and Richard Bright decided to use his own resources and unique methods, so in the end, that left Paul Thomson and me as the "Omega Team" with no official Turkish "contact" or permission—or money! I had never met Paul, so we communicated briefly by e-mail and decided, eventually, to attempt to meet in Dogubayazit at the end of August.

Paul had been on Mt. Ararat before in both 1992 and 1993. In 1992 he had traveled alone to Turkey, and without a team, or permits, or any advance reconnaissance, simply decided to walk across the valley by night and begin climbing. Unfortunately, he was soon thereafter apprehended by members of the Kurdish terrorist organization, the PKK. They apparently decided that he was of little value to their cause, so within a couple of days they let him go with an admonition against returning. Nevertheless, the following year he returned, and was promptly captured again. The second time around proved to be quite an ordeal because the PKK decided that Paul, along with a couple of other foreign captives, might make for good publicity for their plight against the Turkish government. This time he was held as a prisoner for thirty-five days. At times the group guarding him was attacked by the Turkish military and had to endure both helicopter attacks and mortar rounds. Several of his captors were killed during his ordeal, though they always managed to protect their prisoners. At one point, during an attack, he realized his captors were not guarding him, so he escaped and, rather than trying to get down the mountain and away, he ran up the mountain to see if he might catch sight of the ark. Naturally, he was recaptured shortly thereafter—without having seen the ark, though he was able to see into the gorge and glimpse enough to whet his appetite for an eventual return.

Eventually, the PKK realized that they had garnered all the publicity they could get through their holding of foreign prisoners so they let him know that they would be releasing him. They refused to just let him climb down on his own because they claimed the military would kill him and publicly blame it on the terrorists. Paul was not sure if he could believe their motives—having been captive for over a month—but, true to their word, they snuck him down the mountain at night and took him to the police station so that he could be released in full public view and in broad daylight. He was then "arrested" by the military and escorted out of the country and told not to return. Still not entirely sure who to believe, he was able to get a copy of the major Turkish newspaper accounts of his "escape" (as told by official government sources). The account reported that he and the other captives had been bravely rescued by daring military commandos in a firefight with the terrorists. The article reported that the PKK was about to kill him, and he was rescued at the cost of the lives of many terrorists. Having been there in person and knowing that the story was a complete fabrication, he angrily wrote to the papers demanding that he be allowed to tell his own story. He was ignored, and consequently became convinced that the PKK version of his fate was probably the more accurate one—the military would simply have killed him and used it as international publicity against the PKK.

Since our "team" (Paul and I) was so tentative and without official backing of any kind, I continued to struggle with whether or not I should even make the effort. Working only part-time, I did not have the financial resources to make the trip on my own. However, I continually prayed about it and decided to make reservations, even though, at the time, I had received little or no funding and had no resources of my own. I called the ticket agent on a Monday morning expecting to hear back that afternoon with price information. Monday happened to be the very day that a devastating earthquake hit western Turkey, and assuming I was no longer going, my travel agent did not bother to return my call.

When I had not heard back by Wednesday morning, I called again to insist that I was, indeed, going. I left the house to teach my summer school algebra class and returned home for a brief lunch. As I walked in the door, the phone was ringing. The call was from some good friends, Elliot and Robyn Rudell, who had arranged for me to come and teach a series on "Creation, Dinosaurs, and Noah's Ark" at their church following my return from Turkey. They were vacationing in Florida and called just to let me know that if my return from Turkey was delayed, it would not cause problems with rescheduling the start of my series. While we were on the phone they asked me if they could help by paying for my round-trip tickets to Turkey! Before my travel agent had even returned my call with price and date information, they provided me with the finances to pay for the ticket!

Each and every year God's provision has been different (in 1998, I had a motorcycle accident and the settlement eventually paid off all the debts I incurred in going to Turkey—I was grateful for the provision, but I did pray that the Lord would not use that method again). Sometimes it's been last minute or even after the fact, but every year I've seen provision like the manna God provided for the Israelites in the wilderness. Once, I asked God why He did not just prompt someone to search for the ark who already had the necessary finances to go and search, and I felt His response was "If you already had the resources, you would not need me." God has used many friends, family members and churches to give sacrificially to enable me to take the trips to Turkey over the years.

1998 and 1999 Searches for the Ark

Almost as soon as I arrived in Dogubayazit I was met on the streets by "tour guides" who knew I wanted to climb Mt. Ararat and were willing to help—for a price! I was told that "permission" was once again available and it could even be "official and legal." In the spring of 1999, the leader of the Kurdish terrorists (the PKK), Ocalan, was captured by the Turkish military, tried, convicted, and subsequently sentenced to death. Consequently, the PKK had somewhat fallen into disarray and most members had either returned quietly to their homes or fled to bordering Iran. As a result, the military presence around Mt. Ararat was more lax, and claims were even being made that governmental permits for climbing the mountain would again be available in the year 2000.

However, the "permits" available in September 1999 were, in reality, bribes paid to certain military officials who, for a price, would supposedly turn their backs and allow certain climbs to "unofficially" take place. A friend of mine, and fellow researcher, had paid a very large sum to the military and was to be granted five days on the mountain. He hired several locals as guides and porters and headed up the western side of the mountain. The military forced him to carry a cell phone as a "precautionary measure" in case of trouble. Unfortunately, the only "trouble" came from carrying the phone at all! After almost three days of climbing and after just reaching the beginning of their search area, the military phoned to say that their allotted time had expired. The military was able to describe what each person on his team was doing at that moment, so he knew they were being very closely monitored and could ill afford to disobey the command to return. In any event, even though they'd paid a large sum of money, no viable search was actually afforded them. After his return, my friend graciously came to our hotel room to warn us not to climb the mountain without paying off the military because of their ability to monitor everything on the mountain. In the end, though, I felt that paying money to disreputable and dishonest people would likely be counterproductive. As the old saying goes, "You get what you pay for."

Paul and I made several contacts in town, each of whom claimed to be able to help, but in the end, it always amounted to the same thing—pay off the military with substantial sums. Neither of us had the kind of money that was being demanded (usually two to three thousand dollars apiece), but even if we had, I don't think either of us wanted to go that route. Eventually, we met a man named Mustafa (over there everyone seems to be named Mustafa, Mehmet, or Ahmet) who claimed to have grown up on the mountain. He said he had known the astronaut Jim Irwin, and had even acted as one of his guides. He told us that if we wanted to climb without official "permission" or without paying bribes to the military, he would take us himself. He offered to provide us with a horse to carry our gear and said he would take us by night to three thousand meters, where we would camp and find water. We would hide out by day and then climb the next night to forty-two hundred meters. After that, we would be on our own and our guide would leave us. Since we were to be paying someone for actual "services provided" (a horse and guide), instead of a bribe for "looking the other way," we both felt better about where the money was going. I kept remembering a prophetic word that was spoken in my home church the day before I left for Turkey (it was not given to me, or spoken by anyone who knew me, but to the church as a whole): "Do not be afraid to step out into the things that God has called you to and put on your heart from the time you were a child... He is the Creator ... and he will lead you to the mountaintop, and he will not allow your foot to slip." How apropos, I thought.

On September 9, 1999, we were packed and ready to climb the mountain, awaiting only final arrangements for meeting Mustafa, and then Paul got sick. In order for me to be able to wait for another couple of days for him to recover, I had to try and change my tickets for my return flights. After what seemed like several frustrating hours on the phone, I was able to do so, but this only gave me an additional three days. If Paul had been unable to climb within a day or two, I would have again been out of time.

1998 and 1999 Searches for the Ark

As it turned out, Paul needed two full days to feel that he would be able to climb after all, but in that time, the weather took a turn for the worse. In fact, the morning of the third day after his illness, we awoke to rain in Dogubayazit and a beautiful rainbow outside our Hotel Ararat window. I was not sure if that was a portent of God's promise or just a sign of bad weather ahead. As it turned out, it was probably both.

That evening, we met Mustafa on the streets and he threw our packs into his van, saying aloud, "I'll take you to Iran." I suppose such posturing was for the sake of any unwanted ears, but most locals loitering nearby undoubtedly already knew what was happening. The day before, however, our hotel had been visited by the "secret police," who took our passports for a time and began asking questions. We were not detained, but for a while we were concerned about not having our passports—especially if we had to leave town quickly. In the end, our passports were returned with no further questions and we were left alone.

After our packs were thrown unceremoniously into the van we were driven some miles outside town, and near the outskirts of a Kurdish village we met another man crossing the field in the dark with a horse. It turned out that Mustafa was not intending to be our personal guide after all—he had arranged for someone named Mehmet to actually take us. This came as an unwelcome surprise to me because Mehmet spoke no English, and it was not the situation that Paul and I thought we had agreed on. One of the very reasons we had agreed to the arrangement in the first place was that we welcomed the idea of having someone along who was both acquainted with the mountain and who spoke enough English to communicate the basics.

The horse was quickly packed with our gear, however, and it seemed pointless to argue, as Mustafa was clearly not dressed to climb or willing to go. As we headed across the field in the pitch-blackness, I wondered about my sanity.

It would not be the last time. I also pondered whether our money had, indeed, been any better spent than on a mere bribe.

As it turned out, we made two major mistakes before we even began our climb. The first was that, because we assumed that Mustafa was climbing with us and, since he knew where to find water, we left the water filter behind to save weight, and we dumped our extra two liters of water. The second mistake we made was to give my tent to the manager of the Hotel Ararat. I had promised to give him my tent when I was ready to leave, but since we were anticipating the arrival of another friend and potential partner bringing us a new tent, I had given mine away early. Our "partner" never arrived from Istanbul with our new tent, and when we realized we would not have a tent in time, I tried to borrow mine back. Unfortunately, it turned out that Salih, the manager, had gone to Iran and no one else knew where the tent was—we had to climb without.

That evening climb qualified as the most difficult climb I have ever made (until the following year's climb!). It was pitch-black without a hint of light. Most of the time it was impossible to even tell where your foot was going to land when you made a step. I tried to follow the horse as closely as I could (at one point, I even tried to hold on to the tail of the horse!), but occasionally the horse would stumble and I was afraid he would step or even fall on me if I stayed too close. Mehmet was sixty-two years old and I remember thinking two things: "Thank God for the horse that slows down Mehmet, and thank God that the horse is carrying our packs to slow it down." In a footrace for a hundred yards I could probably have outrun Mehmet, but climbing the mountain—at night!—it was no contest, and if it had not been for the overloaded horse, I'd have been completely lost after the first hundred yards. The fully loaded horse was the only thing that kept Mehmet slow enough for me to maintain pace. It was impossible to tell where we were going; impossible to pick out landmarks or any semblance of a trail. I'm not sure if it dawned on me at the time how difficult that would make finding a return route with no guide or recognizable landmarks. I think we climbed until about 1:00 a.m.—up and down, over rocks, and into and out of gullies. I skinned and bruised both knees and tripped more times than I care to remember. If it were not for my hiking staff, I do not know how I would have survived. Paul somehow managed without a staff, but I consoled myself with the fact that he was at least sixteen years my junior—and a more experienced mountaineer!

Somewhere around 1:00 in the morning it seemed as if the horse had done enough and Mehmet pointed to the side of the hill and simply said, "We camp." It was not exactly a level spot (I felt lucky to have not rolled back down the mountain) and I don't know if I slept or not, but it did feel good to be off my feet for a few hours.

Before 5:00 a.m., Mehmet "woke" us as he was packing his horse to leave. Unfortunately, he was headed back down the mountain and not continuing as our guide. He simply pointed up the mountain and said one word, "Direct," and then he left. That meant that we were left without a guide and without a horse to carry our packs—and we were running low on water. We had also been warned several times not to climb in daylight because the military would see us (Mehmet even gave us an oral rendition of mortar rounds being sent in …). Unfortunately, we now had no choice. It was climb and find water or face the next twenty-four hours—and possibly longer—without.

We did our best to stay in ravines where there was some cover from roving eyes, but it was simply not possible if we were to climb "direct" toward the presumed campsite and water. We desperately needed to find both a campsite and water, so we kept climbing. The weather, which had cleared in the evening, also seemed headed for worse. Ominous gray clouds were beginning to build in the south. By that time we had begun to drink from the puddles of water we occasionally found sitting in the cleft of rocks. Sometimes it was none too pleasant, and the surface had to be cleared of detritus of various kinds, but it beat the alternative—no water at all!

By midafternoon we reached a fairly level area and a small enclosure formed by a large rock on one side, and a few piled-up rocks on the other. Since we had only a small tarp for cover, we spent some time trying to use rocks and tie-downs to make as much of a "tent" as we could. It was decidedly uncomfortable and barely large enough for both of us to slither into, but it accorded some measure of relief from the wind and rain and even the sleet that had begun to pelt us. There was no room to stretch out, and getting anything from the packs was virtually out of the question. Eating was decidedly a nightmare. My sleeping position was situated over and around two rather obtrusive boulders, and I do not believe I actually slept more than an hour all night. At one point, I remember looking at my watch and thinking it read 2:50 a.m. I was horrified that I would have to endure at least another three hours before the light of dawn. A few minutes later when I glanced again, I realized that I had inadvertently reversed the hands in the dark and it was actually only 10:10 p.m.! I almost wept.

Our campsite in the rocks—storm clouds are gathering. [Author's photo, September 1999.]

1998 and 1999 Searches for the Ark

We had decided to spend the remainder of the afternoon trying to nap, in hopes that the weather would let up and we could then climb at night to the forty-two-hundred-meter camp (assuming we could find it!). The weather did not cooperate; it got worse. It rained and sleeted and the wind blew relentlessly all night long. I know because I was awake the entire time.

The weather at daylight was no better and showed no signs of letting up. The snow level had come down nearly to where we were camped (somewhere around eleven thousand feet). The good news was that now there were enough puddles left in the rocks that I was able to collect almost two liters of water. It was now Monday morning and my bus would be leaving on Wednesday morning, and my plane on Wednesday afternoon. I realized that if we waited another day for the weather to clear before resuming our climb, by the time we reached the next camp I would have to turn directly around and make the hike back to town (in one day) in order to make my bus and flight connections. In hindsight, it was probably foolish to have tried the climb at all with so little time to spare, but it was all I had. And after three trips to eastern Turkey, it was the first time that I had actually set foot on Ararat.

Paul had another seven or eight days left before he had to make his flight connections, so he decided he would try to wait out the storm and at least be able to do some exploration on his own. (Addendum: I did not find out for several weeks that Paul had waited out the storm for a couple of days, then made his way up the mountain and found the "high camp" at forty-two hundred meters. He came down several days later without having been able to search the gorge as originally planned and eventually delayed his return to New Zealand in order to make a separate climbing attempt with Richard Bright.)

Most of the food supplies I left with Paul. I took one liter of water with me—freshly gathered from the puddles on the rocks—then repacked my pack, and began making my way down the mountain. In the morning hours I had the cover of rain and clouds, for which I was grateful. I was still cognizant of the danger of being seen by either the military or shepherds, especially at the lower elevations. Consequently, I tried to stay in ravines and between hills as much as I could, even though it often made for very difficult climbing conditions (climbing over boulders and through loose rocks).

About midday, I decided to have a swig of water and a little beef jerky for lunch, so I took off my pack and sat down on the hillside. I soon became aware of movement on the adjacent hilltop and realized there were sheep grazing. The presence of sheep means shepherds, and shepherds mean dogs, and both shepherds and dogs mean danger—though each of a different kind. The shepherds are used by the military as informants, and the dogs are used by shepherds as killers to protect the sheep from any intruders. I knew I was not a dangerous intruder, but I was also quite sure that the dogs did not know that. Everyone seems to have their own story about encounters with either the vicious guardian dogs—or their wild counterparts—on Ararat, and none are too pleasant.

In any case, as I was cautiously putting my pack back on, I began to hear dogs barking. At first I did not realize they were barking at me, but I began to make haste nonetheless. It did not take long for me to realize that the sounds were getting closer and I looked up to see three very large and very black dogs headed for me at full speed. I started trying to run down the mountain, but carrying seventy pounds on my back, I quickly realized I would never outrun the dogs. In addition, coming down had caused severe damage to my big toenail and it was excruciating to run (the toenail eventually fell off). Initially, as I was preparing to make my stand, I did not want to seriously injure any of the dogs (though it was a decidedly better option than being seriously injured by them), because I hoped to keep from antagonizing the shepherd further. However, as I waved toward the distant shepherd, trying to show him that I meant no harm, I realized that a fourth dog was now headed my way—the shepherd had either intentionally directed another dog my way, or done nothing to prevent its coming after me. I gathered as many loose rocks as I could find and waited a few seconds until the first dog was snarling and charging just ten feet away. Luckily my aim was good—it was only much later that I was able to reflect gratefully on the years of playing shortstop!—and he both fell and yelped loudly enough to bring the other dogs sliding to a halt. I had a spear at the end of my hiking stick just in case it came down to hand-to-mouth combat, but I hoped they would not get that close. Luckily—or providentially—the dogs did not separate and come at me from several directions at once, so I was able to keep backing down the mountain and throwing rocks at the dogs as fast as I could. I am not sure how long this continued, but it did not end soon enough for me—it seemed like hours, though it was probably only minutes. I'm not sure what I had expected to encounter on the mountain, but my "fear" of the military now seemed rather benign compared to my newfound desire to avoid shepherds and their dogs!

After that encounter, I took out my binoculars and tried to identify herds of sheep far enough in advance that I could avoid them entirely. Unfortunately, this often meant taking a circuitous route and extending the length of my trek considerably. Late in the afternoon, I found a shady spot and lay down to rest for a while. I considered waiting for nightfall in order to finish my trip in the dark, but I knew so little of where I was—in fact, the only thing I knew about my location on the mountain was that I could usually identify down—that I decided it was better to risk detection in the daylight than risk walking off a cliff in the night. I also realized that awaiting nightfall would potentially mean several more hours without water.

As I neared the bottom, I was hailed by a young boy on an adjacent hilltop. I waved and continued walking—fervently hoping he had no dogs. Much to my chagrin, however, he came running down the hillside to meet me. By this time I had been completely without water for several hours, so I hoped I could at least persuade him to offer me some water—"su" in Turkish. I'm not sure if he understood me, but, either way, he ignored my requests. He seemed determined to let me know that I was forbidden to be there (as best I could guess by his gestures), even though he began walking with me and even offered to carry my pack. Though it was a welcome relief to lighten my load for a mile or so, the pack was almost as big as my newfound companion. When we got to the plain, I tried to steer our path toward the highway a couple of miles away. I hoped that I could get a ride from a passing bus or truck for the ten or so miles back to Dogubayazit. My traveling companion seemed determined to head toward a Kurdish village instead, and he seemed to indicate that I could catch a bus there ("autobus"). I prayed continuously for wisdom, direction, and discernment. It occurred to me that it would be better for me to be seen crossing the open plain with a shepherd boy who belonged there, than to be seen by myself. It also came to me that if I went my own way in spite of his best efforts to help me, he might run off and get "help" of a different kind—and the thought that his rejected help might eventually include dogs was too much to bear.

We walked for several miles toward the distant Kurdish village and by the time we reached it, it was evening and dark. Instead of a bus stop, however, my "helper" took me straight to the back of a house where he "surrendered" me to an elderly man—probably the village patriarch. This new threat immediately began yelling something at me in Kurdish (which I presumed to be words to the effect that I was forbidden to be on the mountain). I smiled and said "good-bye" and miserably shouldered my pack to begin walking back toward the highway. It had now been six or seven hours of climbing with no water and I was in desperate need of a drink. I could hear the yells recede behind me and, though I refused to look back or acknowledge them, I was grateful that the sounds did not pursue me. I kept dreading the sound of vicious barking but it never materialized.

I realized that I now had several more miles of hiking across the fields in the dark with no guarantee of getting a

1998 and 1999 Searches for the Ark

ride once I reached the highway. Besides which, the two persons behind me might very well decide to get help and come after me. But after walking maybe a hundred yards, I encountered a lone man walking across the field in the dark. As best I could, I asked if there was a bus stop in town ("autobus"). "No autobus," he said, and kept walking. It then occurred to me to ask for a taxi to Dogubayazit. With that, he stopped and said, "Me taxi." When I asked in Turkish, "How much?" he responded with, "Five million lira" (about eleven dollars at that time). I showed him that in my wallet I had just five million lira. Unfortunately, he quickly grabbed the money and bade me follow him (a note to future travelers, never pay before the ride...)

On the way to the village, I asked for "su" and was rewarded with a grunt and a nod. Eventually, right on the outskirts of the village, my guide stopped and pointed to a gusher of water coming up through the ground in the middle of the dirt road. Thoughts of giardia and sheep dung filled my mind, but my lips were too parched and my body too dehydrated to think for long. I took off my pack and dropped to my knees and began guzzling water. I prayed for grace and I drank my fill. Thankfully, though throughout the trip I had been drinking from stagnant puddles and now from unfiltered water in the midst of a shepherd's village, the Lord was indeed gracious, and I never experienced even a hint of sickness or diarrhea.

Back in the village from which I had fled only minutes before, there was a lone car waiting on the dirt road and already full of passengers—clearly not a "taxi." Somehow they made room for me and put my pack in the trunk, but the person who had taken my money did not accompany us. As we drove away, I tried to explain that I had already paid the missing benefactor five million lira. No one seemed concerned (but me). However, after a few miles, most of the other passengers were let out, drivers were changed, and someone new proceeded to drive me back to town and my hotel sanctuary. As soon as we arrived, the new driver now demanded that I pay him the agreed upon fare of five million lira. I attempted to explain that I had already paid all my money to his friend and I showed him my empty wallet. He followed me angrily into the hotel and began demanding of the night manager that I pay the "taxi" fare. Explanations were useless as neither of them spoke English, and I spoke no Kurdish, so finally, in exasperation I told him to call the police and I took my pack to my room. There, I took a lengthy shower and changed into clean clothes and then cautiously headed out to find a place to buy some clean, cold water—and an Orange Fanta! Luckily, the driver had apparently realized the futility of pursuing his scam further and he was gone by the time I came down. I do hope he obtained his fair share from the man who took it. A word to the wise in Turkey: establish the cost of services in advance, and pay only upon the completion of same. That is especially true for seekers of the lost Ark of Noah.

The following day I encountered Mustafa on the streets of Dogubayazit and he seemed genuinely happy to see me alive. We found an empty upstairs restaurant room to have tea and to debrief. At that time, I was quite angry because I felt he had misled us. He was supposed to have been with us for two nights of climbing—up to forty-two hundred meters—but instead he left us with an alternate who spoke no English and who abandoned us after the first night. He seemed quite surprised at my frustration. According to his version, Mehmet left us because we "were very strong climbers and did not need his continued help." ("Ha," I thought, "he must never have seen me holding to the tail of the horse!") As we ironed out our differences, he suddenly said, "You are here to find Noah's Ark." When I concurred, he told me that he had seen it once in his life, and that he had never told anyone. He drew a picture of a dark, rectangular object sitting on a ledge. From below, he said, it could not be seen. From above, most people had overlooked it because of an overhang. Approach is very difficult, he maintained, and requires ice equipment and at least two people. He said he felt that I was the "one good people" he had been waiting for, and that I had the ability to speak to the media, which he felt he did not. If I would go with him, all he wanted in return was for me to tell the world that he had been with me.

However, he maintained that even though the government would allow legal climbs of the mountain in the year 2000 (in that respect, at least, he was correct), the military would never allow free access to the area where this "structure" is located (whether or not any of the other things he told me were true, I believe this latter piece of information to be true based on years of personal observations and experience). Any access to the location of the ark that he claimed to have seen—only once—would have to be made without permission and in secret. Consequently, he said, we would need to go alone, just the two of us, because any more than that would be too many to hide or keep secret. The picture that he drew for me and the information he gave me convinced me that he had either seen something, or heard enough stories from those who had, in order to make his story convincing. The real question was whether or not, in a year's time, he would follow through on his promise. But I decided it was enough of an opening that it was worth pursuing. So I began laying my plans for the next year's expedition: 2000.

Chapter 6

THE BIBLICAL ACCOUNT OF THE FLOOD

In secular institutions and scientific literature, the story of a global flood and Noah's Ark is one of the most ridiculed stories of Scripture, second perhaps only to the notion that the earth is approximately 6000 years old. Evolutionist Stephen Jay Gould has pointed out that virtually every modern text on geology begins by ridiculing the concepts of a six thousand-year-old earth and a global flood. He recognizes that Archbishop Ussher, who used the biblical genealogies to arrive at a date for creation of 4004 BC, had impeccable scholarship and brilliant credentials, and yet Gould argues that the fatal flaw in Ussher's reasoning was not in his system or scholarship, but the fact that he used the Bible at all.[1] As Gould states: "Today we rightly reject a cardinal premise of that methodology—belief in biblical inerrance ..."[2] In other words, Gould's premise is that the Bible is false and conclusions based upon its history—such as the occurrence of a global flood—are untenable. In other words, belief in evolution and the concept of billions of years of time, is built upon a foundation that rejects biblical history. That given, it is difficult to comprehend why many Christians are apparently anxious to insert into the Bible a "history" that is based on the premise that the Bible is false.

I believe that the amount of ridicule heaped on these two stories—a relatively recent creation in six literal days, and the global flood of Noah—establish them as spiritual weather vanes that directly point to their significance. Unfortunately, it seems that many Christians have not recognized the attack to be—as even Gould did—in reality, an attack on the very foundation of Scripture. Faced with attack and ridicule, many Christian scientists and theologians have backpedaled, saying, "Well, the Bible does

The Biblical Account of the Flood

not really require the flood to be global, nor the days of creation to have been literal. These are just allegorical stories." The resulting compromise is threatening the very foundations of Christianity.

Perhaps the reason the flood account is so ridiculed is that the implications are so overwhelming if it were true. There are at least three main repercussions of establishing the occurrence of a global flood a few thousand years ago:

(1) If there was a global flood less than five thousand years ago, then most of the major geological features of the earth were produced very rapidly in the processes and aftermath of that event. As Dr. Derek Ager pointed out in *The Nature of the Stratigraphical Record*, "The hurricane, the flood or the tsunami may do more in an hour or a day than the ordinary processes of nature have achieved in a thousand years."[3]

Artist's rendition of the ark enduring the waves of the flood. [Used by permission, courtesy Tim Lovett, Answers in Genesis, worldwideflood.com, and www.answersingenesis.org]

Even small, local floods can produce devastating damage—a year-long, global flood would have produced or reshaped the continents, formed mountains, and filled the oceans of the earth. Instead of forming by gradual processes over millions of years of time, geological features like the Grand Canyon would have formed virtually overnight by rapid, cataclysmic deposits of sediment and flooding.

(2) Basically, the complete fossil record, which includes the dinosaurs, was produced in that one event. Which necessarily means that the entire concept of vast eras of time—which is the fundamental requirement of evolution theory—is completely eliminated. Nevertheless, many people who believe in the Bible as history are taken aback when they are first informed that dinosaurs were on Noah's Ark. Why is that even a surprise? Media, educational, and most "scientific" venues have so popularized stories of dinosaurs with the invariable attachment of millions upon millions of years, that such pronouncements have not even been examined in the light of Scripture, let alone rejected. But the truth is that only a cataclysmic event—or series of events—involving massive amounts of water could have produced the fossil record. Fossils do not form any other way and, in fact, cannot be demonstrated to be forming anywhere on earth today. Dinosaur fossils simply cannot be sifted out of the fossil record while we try to maintain that they died millions of years before the fossils underneath them were formed. In other words, the fossil record as a whole cannot be seen as evidence of a global flood without including the dinosaurs in that event—both living in the world, and entering into the ark. The dinosaurs are part and parcel of the fossil record and usually occupy the uppermost layers thereof.

In Genesis 1:25 it says that "God made the 'wild animals' according to their kinds (this would include many of the dinosaurs), the 'livestock' according to their kinds, and all the creatures that move along the ground according to their kinds (reptiles, etc.)." In Genesis 6:14 notice that the list of animals taken onto the ark with Noah is identical to the list of created kinds: "They had with them every wild animal according to its kind, all livestock according to their kinds, every creature that moves along the ground according to its kind and every bird according to its kind." Understanding the fact that there was a global flood, and the fact that many historical sources—some quite recent—record human encounters with dinosaurs (even though they may not have been named as "dinosaurs") clearly demonstrates that they had to have been included on the ark. On what biblical grounds could dinosaurs be excluded from the ark: size or diet? Then on what grounds could elephants or lions be included? Most dinosaurs were far smaller than full-grown elephants, and none could be more carnivorous than lions (but see chapter "The Scriptural Advent of Animal Carnivorousness"). According to Hugh Ross, it was only "domesticated animals within the sphere of man's influence" that were taken on the ark.[4] Yet there is no hint of such an interpretation in the biblical account of the flood. In fact, the inventory of Genesis 6:14—those animals taken onto the ark—specifically coincides with the inventory of Genesis chapter 1 that lists every air-breathing land animal and bird created.

(3) The third implication is that, not only does the Bible have to be seriously understood as true history, but it demonstrates that there must be a God who communicates with man. Noah did not build the ark on a whim thinking, "It never has rained before; in fact, I don't know what rain is, but who knows, maybe in a hundred years or so, I'll need a boat big enough to save my family and all the animals of the earth."[5] Only the Creator Himself could have given Noah enough notice of a global flood along with the necessary dimensions to enable him to build such a vessel.

The question that needs to be addressed is not "Is a global flood scientifically feasible?"—otherwise, the crossing of the Red Sea, the plagues, healing miracles, the Incarnation, the Resurrection, and on and on, would also need to be rejected—but "What does Scripture actually say about such an event?" That should be the entire crux of the matter. Tragically, there are many Christian scientists and "theologians," who have tried to reinterpret

Scripture in light of modern-day "scientific" prognostications, but the ultimate result is always compromise and contradiction. When one portion of Scripture is twisted to somehow reconcile with human pronouncements, another portion always comes undone. In the end, Scripture is made secondary or irrelevant.

The reality is that the biblical account of Noah's Flood is not in any way obscure or ambiguous: it is only modern misconceptions that have clouded the issue. Genesis 6:7 begins with God (not Noah's perspective) saying "I will wipe mankind ... from the face of the earth—men and animals, and creatures that move along the ground, and the birds of the air." Even a flood large enough to cover the entire Mesopotamian Valley would have had little or no effect on the bird population. Virtually all birds would be able to fly above it far enough and long enough to find a new, drier home.

Genesis 6:13 reiterates God's pronouncement: "I am surely going to destroy both (people) and the earth." There are undoubtedly better ways that God could have announced that he only intended to wipe out the region of the earth where men lived, but no better way to say he intended to destroy the whole earth.

In Genesis 6:17, God reiterates (not Noah), "I am going to bring floodwaters on the earth and destroy all life under the heavens, every creature on earth that has the breath of life in it [note: this is not said of fish or sea creatures]. Everything on earth will perish." Three different times in three different ways, God emphasizes that every living creature on earth will perish. This is not a pronouncement from Noah's perspective, but from the perspective of the Creator of the Universe. So the question is, if God meant that a "local" flood was going to destroy the Mesopotamian Valley, could he not have found a way to say it that did not deceive myriad generations of Jews, Christians, and secular historians alike?

One of the arguments given in secular circles against the idea of a global flood is that Noah could not possibly have gone to every corner of the world to collect the

The Biblical Account of the Flood

kinds of animals that live in different climate zones and ecological niches. But notice in Genesis 6:20 that God tells Noah the animals will come to him. Again in 7:8 it says that pairs of animals came to Noah. In the fossil record (and in places like the La Brea Tar Pits in California) animals are found from what would now be considered different climate and ecological zones all jumbled together—even those from what are considered by the paleontologist to be vastly different eras of time. Furthermore, one feature of some current animals that has remained scientifically inexplicable is their ability to migrate vast distances—evidently, it is not beyond the ability of the Creator to direct animals to the place they need to be.

One of the questions that must be answered by any advocate of a nonglobal flood would have to be "Why build an ark for animals at all?" Hugh Ross, in his video entitled *The Universal Flood*, makes the statement that "It wasn't that God was concerned about the preservation of these creatures, given that the flood was simply extensive enough to cover the area of the reprobation of mankind, ... these animals could simply have wandered back into the Mesopotamian region."[6] He goes on to say that the reason God had Noah bring animals at all is simply so that Noah and his family could have their immediate food and clothing needs met following the flood. But contrast Dr. Ross's pronouncements with the words of the Creator himself in Genesis 7:2-3: "Take with you seven of every clean animal, a male and its mate [likely meaning seven pairs, or fourteen of each[7]], and two of every kind of unclean animal, a male and its mate, and also seven of every kind of bird, male and female, *to keep their various kinds alive throughout the earth* [italics added]."

God's pronouncement is diametrically opposed to the statements of Dr. Ross. If the same animals existed everywhere else on earth—as every local flood scenario requires—then this verse in utter nonsense. In his video, Hugh Ross admits that within a decade all the animals from outside this region would simply have wandered back into the Mesopotamian Valley and repopulated it.

According to Ross, then, God had Noah take with him on the ark the domestic animals that he and his family would need for food and clothing in the immediate aftermath of the flood. But how many years would it take to build up a large enough herd to begin eating from it without eliminating it? Two? Five? Ten? By this time, even according to Hugh Ross, all the animals would have wandered back and repopulated this region anyway. In fact, it is a completely arbitrary number to say it would have taken ten years. It could have happened in two! Or less. Consequently, according to any "local flood" scenario, God required Noah to spend untold resources and (perhaps) hundreds of years of intensive labor to save maybe a year or two at the other end. And he did it all in order to save a few animals that existed everywhere else on earth and would have wandered back into this region in a couple of years anyway.

Furthermore, the text insists that the "wild animals, the birds, and the animals that move along the ground" (reptiles, etc.) are included in God's salvation orders to Noah. Most local flood reconstructions do not even attempt to answer why any animals besides those who met Noah's immediate needs would have been included at all. Why would there have been any need to include animals that were not essential to Noah's physical needs immediately after the flood? According to any local flood scenario it would be counterproductive to include carnivores that would likely need to prey on the very animals that Noah and his family needed for their sustenance after the flood. The carnivores leaving the ark would have quickly become a threat to the very animals Noah needed for his family's survival (but see chapter "The Scriptural Advent of Animal Carnivorousness").

Perhaps a more obvious question would be "If the flood was indeed simply extensive enough to cover 'the area of man's reprobation,' as Hugh Ross claims, why did Noah not just move?" Certainly if the animals could simply have wandered back into this area within a few short years after the flood, Noah could have moved to a safe location (where all the non-ark animals safely survived the flood) within a very short time. In fact, Noah could have gone to where the animals dwelt in safety, without building any boat, and he could have done it with a lot less luggage! Even if one tries to argue that Noah would have had to sail to a safe haven, he could have done it with a much smaller vessel and simply met up with the animals he needed in his new location. Any "local flood" scenario makes a mockery of the story of Noah building an ark at all; it is an absurd attempt to twist Scripture to fit with the pronouncements of secular scientists—pronouncements that were intended by their very nature to preclude the historicity of the scriptural account in the first place.

In Genesis 7:11 it states that "all the springs of the great deep burst forth, and the floodgates of the heavens were opened"—the precipitating event of the flood was the eruption of the fountains of the great deep. Some who have adhered to a "local flood" scenario have tried to argue that there could not be enough water in the atmosphere to cover all the "high mountains of the earth" with the waters of a flood (why this would be a problem for the Creator of water itself as well as the earth is somewhat lost on me), but Scripture makes clear that the waters were not simply those that existed in the atmosphere.

There are two basic questions that secular scientists and Christians alike have demanded of the flood event: (1) Where did all the water come from? (2) Where did all the water go after the flood? These may be interesting questions to discuss, but realize that they are not necessary precursors to understanding the meaning of the text. It is the text that enables us to properly understand nature, not science that enables us to reinterpret the text! Let me illustrate it this way: Do you believe that Jesus fed five thousand men—not counting the women and children—on five loaves and two fishes? If not, then no naturalistic explanation will suffice to redeem the story. If so, where did the rest of the fish come from? Was it a mirage that deceived people into thinking they were eating and being filled? Or did the fish come from naturally born fish existing in the Sea of Galilee to the hands of Jesus (filleted and cooked along the way) in some kind of time/space warp? Or is he the Creator who simply made as many fish as he needed? Already cooked, cleaned, and ready to eat. Having insufficient water to cover the world's highest mountains is no more of a problem to the meaning of the text than having enough fish is a problem to the Feeder of the multitudes. So, you can choose to accept or reject the clear meaning of the text, but you cannot rewrite it to fit your own notions of "science."

As an aside, one could ask, how old did the fish appear that Jesus used to feed the multitude? If they were created in an instant, but appeared to be several years old, would this not be a deception? Many people have attempted to use a similar argument to say that since we seem to see light from stars that are apparently billions of light-years away, there must have been billions of years of time available, or this would be a deception on God's part. But in the story of the feeding of the multitude, there can be no case for deception since we are *told* it was a miracle—done literally in front of thousands. In the case of starlight, ignoring for the moment the fact that there are innumerable untestable assumptions that go into declaring that stars are billions of light-years away, there can be no "deception" because we are *told* that the stars were created on the fourth day of creation. It is written in the creation account told by the Creator himself. As in the case of turning water into wine, the process is recorded, not as a natural process, but as a supernatural act of creation! So, if ultimately the scientists have no explanation for where the water of the flood came from—or where it went afterward—it is meaningless to the interpretation of Scripture. If you can reinterpret this part of Scripture to make it palatable to the modern scientist, you will ultimately create a text that is devoid of any real meaning.

That is not to say that there is no evidence for either where the water came from or where it went. It simply means that if we, as finite man, do not have all the answers, we must look to the text for answers and not to man's interpretation of nature. Genesis 7:11 seems to indicate that the water was not created for this event but already existed on earth—or *in* the earth. The very crust of the earth broke open and the "fountains of the great deep" burst forth. Geologist Walt Brown believes that the pre-flood crust rode upon a layer of water and the precipitating event of the flood was the breaking of the crust, which then expelled water at tremendous rates and volumes high into the atmosphere.[8] Some also believe there was a water-vapor canopy around the earth that helped to maintain a worldwide temperate climate. Textually, this seems to be indicated by the mention of the separation of the waters above the "expanse" (sky) from the waters below (Genesis 1:6-8). However, physically there could not have been enough water in this canopy to produce forty days of rain and continued rising waters for 150 days. Consequently, it is likely that the majority of the waters now contained in the oceans was once under the earth's crust.

In Genesis 9:15 God promises Noah that "never again will the waters become a flood to destroy all life." Once again, if the flood was only local, then this promise must be seen to be misleading or downright false, since there have been many devastating and even cataclysmic floods on the earth since the time of Noah. Nevertheless, this does provide evidence for the disposition of the water after the flood: the ocean basins were formed to contain the water in such a way that a worldwide flood cannot physically happen again. In fact, Psalm 104:6-9 states: "You covered (the earth) with the deep as with a garment; the waters stood above the mountains. But at your rebuke the waters fled, ... they flowed over the mountains, they went down into the valleys [the ocean beds], to the place you assigned for them. You set a boundary they cannot cross; *never again will they cover the earth* [italics added]." Secular scientists and even Christians like Hugh Ross make the mistake of looking at the earth as it is configured today—instead of looking at the Word of God—and arguing backward that the waters in the oceans as they exist could not rise and cover the mountains. And the problem is that they are entirely correct, but entirely irrelevant as well! The oceans today exist as an effect of the flood, and the Scripture specifically says that the earth was reconfigured in such a way that a global flood *could never happen again*. It is God's promise to Noah that a global flood can never occur again and scientists who make that observation are only confirming the biblical account as accurate.

But the real keys to understanding the nature of the flood lie in Genesis 7:19-20. These verses say that the waters "rose greatly on the earth, and *all the high mountains under the entire heavens were covered*. The waters rose and covered the mountains to a depth of more than twenty feet [note: the Hebrew text says "fifteen cubits" which equals 22.5-30 feet]." [Italics added.]

These verses alone are enough to demonstrate that the text cannot mean anything other than a global flood. The Hebrew phrase "under the entire heavens" is a redundancy that is reiterated from God's earlier pronouncements and it cannot mean anything other than the entire sphere of the earth. The phrase "all the high mountains" is so significant that those opposed to the concept of a global flood have gone to great lengths to avoid it or refute its clear meaning. In 1990 for example, Hugh Ross tried to deal with it by eliminating it from the Bible entirely. In a taped lecture series he stated, "The word 'high' [Hebrew *gaboah*] is not in the original. It's not there... It's not in the Hebrew."[9] Unfortunately for his claim, the word is in every available Hebrew manuscript as well as every modern English translation. When confronted with his error, he issued no retraction, he simply deleted those few sentences from the tape and continued selling it. He now deals with the verse by saying that the word for high—*gaboah*—is a relative term that was written from Noah's perspective to mean that "the hills he could see were covered." However, it seems self evident that if the meaning of the text were not clear, Ross would never have seen a reason for deleting the word "high" in the first place. The fact that he tried to do so indicates clearly that he was attempting to force the text to agree with his position, not honestly trying to discern the real meaning of the text. He also ignores the phrase "under the highest heavens" completely. And once again, the truth is that if these verses do not mean a global flood, then not only were myriads of generations deceived, but apparently the concept of a global flood is not a linguistic possibility in the Hebrew language.

In verses 21-23 (chapter 7), the text reiterates and reemphasizes in at least three different ways that every animal with the breath of life in its nostrils on the face of the earth would be destroyed. And if that was not enough, in verse 23 it says that on the whole earth, only Noah and those left with him survived. That clearly includes the animals.

Ultimately, I do not believe God is going to continue to tolerate the current Church's rewriting of Scripture to make it more palatable to the modern so-called scientific mind. In his video, "The Universal Flood," Ross states, "We shouldn't look for the Ark of Noah. I don't think it will ever be found."[10] Then he goes on to say that, even if it could be still around, it could not be on the top of, or anywhere near, Mt. Ararat. He says that by the evidence of the text, and the evidence of geology, "we see only evidence for a flood throughout this region of Mesopotamia; [it] doesn't extend as far as Mt. Ararat."[11] In fact he goes on to say, "So really, the worst place to try to find the ark would be anywhere near Ararat itself. It's at an elevation far too high [in other words, the waters could not have arisen that high or it would have to have been a global flood!]... *It can't be there.* And yet a lot of money is being spent by Christians looking for something that can't possibly be there [italics added]."[12]

Clearly, the flip side of this statement is that, if the ark is found on Mt. Ararat, the flood was global, the fossil record is a record of that event, all evidence of a vast era of time is eliminated from geology and from the Scriptures, and any attempts to rewrite Scripture to help it coincide with modern science are patently false and ultimately undermining the veracity of Scripture and the foundation of Christianity. Indeed, when Noah's Ark is ultimately revealed to still exist on the top of a seventeen-thousand-foot mountain, every alternative to a global flood will be clearly exposed as foolishness.

The evidence at this point for the continued existence of Noah's Ark on top of Mount Ararat in eastern Turkey cannot be said to be incontrovertible. On the other hand, it is more than just intriguing, and taken collectively, it is hard to discount. Numerous eyewitness accounts emerged in the twentieth century claiming the remains of Noah's Ark are still high on Mt. Ararat. Many of them have been thoroughly reported elsewhere,[13] so I will just summarize two of the more persuasive accounts.

Ed Davis was an American stationed in Iran with the American military in World War II. According to Ed's account he developed a friendship with a young man hired by the military as a driver. One day Ed pointed to a snowcapped mountain and asked his friend if that was not Mt. Ararat, which the biblical account referred to as the resting place of the ark. His friend, Badi Abas, replied that their name for the mountain was "Agri Dagh," and it was indeed the location of the ark. There are researchers and authors (most notably Bob Cornuke—author of *In Search of the Lost Mountains of Noah*—who has also searched on Ararat for the ark) who believe that Ed Davis may have been referencing a mountain in Iran or elsewhere in Turkey rather than the current Turkish Mt. Ararat. However, it would be difficult to use part of Ed Davis's account and delete the parts that are inconvenient to an alternative interpretation. The current Mt. Ararat in eastern Turkey is referred to by the Kurdish shepherds as "Agri Dagh," and this is not known to be true of any other mountain in the area or in Iran.

According to Ed's account, the young shepherd told him that his family was privy to the location of the ark's remains, but that it was not visible every year. If it were to become visible that summer, Badi told Ed, he might be permitted to travel with the Abas family to see it. Sometime later, Badi's father, Abas Abas, the patriarch of the

The Biblical Account of the Flood

clan, showed up to say that the ark was now visible and they were going to see it. Ed somehow finagled permission from his commanding officer to take a leave of absence and tag along. He claims that they drove by truck for eight to nine hours, and that for some of that time they were traveling along the Russian border. Even a cursory glance at a map of Turkey and Iran makes it clear that the only way to have driven along the Russian border from Ed's station in Iran was to have headed for eastern Turkey and the current Mt. Ararat. Ark researcher and author Bob Cornuke avoids this conflict in his own interpretation of Ed's account by reasoning that Ed saw Russian soldiers in northern Iran and driving along the Caspian Sea might have been considered the "Russian border."[14] This seems highly unlikely, since Ed was stationed with the American military and the lines of demarcation between both armies and countries would necessarily have been well known to all and sundry. In any case, Ed never mentioned seeing Russian soldiers as they drove, but he clearly did say that they drove for some length of time "along the Russian border."

Cornuke further attempts to bolster his identification of Mt. Sabalon in Iran as the biblical Mt. Ararat (and location of Ed Davis's sighting) by citing Genesis 11:2, which states that after the floodwaters subsided, "Noah's descendants journeyed 'from the east' to settle in the land of Shinar."[15] (Note: Since the publication of his book, Cornuke has withdrawn his claim of Mt. Sabalon as the "Ararat" of Davis, and in its place submitted the Iranian Mt. Suleiman instead.) Since many biblical scholars recognize "Shinar" as the area of modern-day Baghdad, Cornuke reasons that the ark could not have landed on Ararat in eastern Turkey, since Noah's descendants could not have then moved "from the east" to get to Shinar. Unfortunately, in order to translate Genesis 11:2 as stating that Noah's descendants traveled "from the east," Cornuke has to rely on the Septuagint version of the Old Testament. The Septuagint is a translation of the Hebrew into Greek and it is demonstrably inferior to the original Hebrew. There are numerous recognizable inaccuracies as well as interpretive insertions that are not true to the original Hebrew. In fact, the NIV renders the same verse (working from the best interpretation of the original Hebrew), "As men moved *eastward*, they found a plain in Shinar and settled there [italics added]." That gives completely the opposite meaning to the verse. In addition, not all scholars concur that "Shinar" was in the area of Baghdad, so it certainly would be premature to forge an identification of the biblical Mt. Ararat on this particular verse, especially when Mt. Sabalon (and now Mt. Suleiman) does not coincide with so many other aspects of Ed Davis's account.

By Ed's account, they spent eight or nine hours driving until they reached the foothills, where they then mounted horses and continued their trek. They stayed in one village (perhaps Ahora Village at the base of the Ahora Gorge) where he was shown artifacts taken from the ark, including what looked like oil lamps, pots for food, and even a cage door that appeared to be petrified. According to the shepherds, after the ark broke apart in the early twentieth century, they began to retrieve artifacts that fell out of the broken sections.

According to Ed's account they were taken by horseback into a large gorge where he was shown sites referred to as "Jacob's Well" and "Jacob's Tomb." Later they stayed in a large cave referred to as a hideout of Lawrence of Arabia. (The sites of Jacob's Well and Tomb are well documented and photographed in the Ahora Gorge of Mt. Ararat.) They spent three difficult days of climbing in cold, dreary, and rainy conditions, staying in caves along the way. In one cave Ed testified that there were ancient and beautiful carvings on the walls. In the late afternoon, as the clouds parted and the sun broke through, he was taken to the edge of a cliff where he could look down into what he called a "horseshoe-shaped" canyon and he saw an enormous wooden ship that was broken apart into at least two sections. He could look into the end of the upper piece and see three levels or decks with broken timbers protruding. Farther down the mountain he saw a second piece that had separated from the first and apparently slid down. The shepherds told him that the ark had been intact until the early part of the twentieth century, when, because of the ice filling it, it broke apart. He was told that he could be taken to the upper piece, by using ropes, the following day. However, during the night, the rain turned to snow, and in the morning, the now-buried ark was no longer visible. Because of the dangerous terrain and the snowfall, they were unable to descend into the canyon containing the ark and they had to begin their descent of the mountain. It became a very difficult and exhausting descent that extended into five days.[16]

Ed has been extensively interviewed by several ark researchers and in the late 1980s he agreed to submit to an extensive lie-detection test. He spent roughly three hours with the analyst and the analyst concurred that he had indeed seen a large wooden vessel on top of a mountain. Ed always believed that he was on Ararat in eastern Turkey but he could not prove that either to himself or researchers. Consequently, there has remained enough room for doubt that researchers like Bob Cornuke have been able to utilize the essential veracity of Ed's account while placing him on various mountains in Iran.

In my estimation, at least three things preclude this possibility: (1) The route he describes of driving for eight to nine hours from Quasvin and driving for several hours "along the Russian border" could not have taken him to Mt. Sabalon or Mt. Suleiman or any other mountain in Iran. (2) His eyewitness description of Jacob's Tomb and Jacob's Well coincide with known locations in the Ahora Gorge of Mt. Ararat in eastern Turkey and have not been identified on any other mountain in Iran or elsewhere. (3) The Iranian shepherds told him that their name for this mountain was "Agri Dagh," the name currently used of Mt. Ararat in eastern Turkey (this was volunteered by Ed, and not solicited by any questioning that could have planted such an idea in his subconscious). There is no known source for calling any mountain in Iran or elsewhere by this name.

Dinosaurs on the Ark

In addition, there are several specific details of Ed's account that I believe I was able to corroborate and uniquely connect to Mt. Ararat on my 2003 trip to this region with Richard Bright (see chapter "The 2003 Search for Noah's Ark).

The second account is that of Vince Will, an airman stationed in Naples, Italy during World War II. Vince later became good friends with my friend and climbing partner Richard Bright. After his time in the service, Vince became a Lutheran minister and the story that he told of Noah's Ark never wavered or varied over the years. As he told it, one day a group of pilots were looking over some photographs in the mess tent. He was shown a photograph that clearly depicted a portion of a large wooden ship jutting from ice and sitting on a ledge of a snow- and ice-covered mountain. He realized with some excitement that they were looking at the remains of Noah's Ark. Upon also seeing the photos, Vince's pilot friend queried about the specific flying route one needed to take

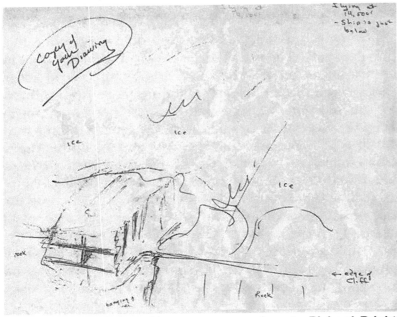

Vince's sketch of the ark's appearance—given to Richard Bright. [Used by permission.]

The Biblical Account of the Flood

in order to see the ark for himself. Shortly thereafter, on their flight to Erivan, Turkey, the pilot flew the plane in a clockwise path around Mt. Ararat. As they flew the route and broke free from the clouds, Vince could look out of the window down to his right and he could see the prow of a huge wooden ship jutting from the ice (see sketch).

Vince slipped into the left side pilot's seat and took over the controls so that the pilot could get a better look out the right side window. Though Vince had a private pilot's license and was capable of keeping the plane aloft, he was not an army air corps pilot on the crew. The pilot then took control from the co-pilot's seat and dipped the right wing for a better view. Vince was then able to get a clear look into the structure. He later told Richard that he had no doubt whatsoever that he had seen the remains of Noah's Ark. He was adamant that what he had seen was not a rock or a cave or even some other type of man-made structure. He was so convinced that he had seen Noah's Ark that he once told Richard that he "would stake his salvation on it."[17]

Ultimately, there is no biblical requirement that the Ark of Noah was preserved for all time. There is no further biblical mention of the continuing existence of the ark after the account of Noah. Yet its current existence and rediscovery could have enormous repercussions. As stated previously, the biblical account of a global flood is neither unclear nor ambiguous. In fact, if the biblical account of the flood was not intended to reveal that there was a global flood between four and five thousand years ago, I believe you must reach three highly implausible conclusions: (1) Noah, himself, a man of righteousness who intimately communed with God, was unclear as to the extent of the flood; he thought that a flood that covered the Mesopotamian Valley actually covered the entire earth because all that he could see of the earth was covered.

(2) Scripture was not only misunderstood, but unable to be understood, for thousands of years until the

advent of modern geology—a discipline that was largely designed by those seeking to eliminate the biblical account as history. Charles Lyell, for example, called the biblical account of a global flood an "incubus" to modern geology. Lyell, who was largely responsible for promoting the theory of the "ice age" was attempting to explain geological features that had actually been formed by the flood, by inventing continental sheets of slow-moving ice instead (see chapter "The Myth of the Ice Age").

(3) There is apparently no way to say within the Hebrew language that there actually was a global flood. In other words, if the language of the text as it exists does not mean there was a global flood, then such a concept seemingly could not be explained in the Hebrew language. For anyone who claims to believe in the veracity of Scripture and yet rejects the global nature of the flood, I would issue a personal challenge: find a better way in the Hebrew language to express that there was a global flood.

There are very few places on the entire earth as remote and inaccessible as Mt. Ararat where it is even possible that such a structure as the ark could remain preserved—and mostly hidden—for thousands of years. Consequently, I believe that God has preserved the ark for such a time as this—when mankind, and even a large portion of the Church, has rejected the foundation of the truth of the Word of God and exchanged it for a lie. II Peter 3:3-7 states, "You must understand that in the last days scoffers will come, scoffing and following their own evil desires. They will say, 'Where is this "coming" he promised? Ever since our fathers died, everything goes on as it has since the beginning ... [Lyell's uniformitarianism!].' But they deliberately forget that long ago by God's word the heavens existed and the earth was formed out of water and by water. By these waters also the world ... was deluged and destroyed. By the same word the present heavens and earth are reserved for fire, being kept for the day of judgment and destruction of ungodly men."

The Biblical Account of the Flood

Everyone will not automatically believe in the veracity of the biblical account of history when the ark is fully revealed once again. There were many among the Jews and Pharisees who, with their own eyes, saw a blind man regain his sight and a lame man walk, and yet did not believe in Jesus. *But many did.* And many will! Mankind is going to be left without excuse in these latter days! God desires that not one would perish.

Endnotes

1 Gould, Stephen Jay. "Fall in the House of Ussher." *Natural History*, Vol. 100 (November 1991): 15-16.
2 Ibid., 16.
3 Ager, Derek. *The Nature of the Stratigraphical Record.* New York: John Wiley & Sons, 1993, 80.
4 Ross, Hugh, Ph.D. Video: *The Universal Flood.* Pasadena, California: Reasons to Believe, 1993.
5 According to Genesis 2:5-6 it had not rained prior to the flood: "For the Lord had not sent rain on the earth ... but a mist came up from the earth and watered the whole surface of the ground." Some commentators contend that these verses do not refer to the entire time before the flood, but if they do not, then God's post-flood promise of a sign in the sky (rainbow) becomes difficult to comprehend. If it had rained before the flood, the rainbow would already have been a well-known and commonplace phenomenon.
6 Ross, op. cit.
7 Though the English text reads "seven of every kind," I believe a better rendition would be "fourteen" or "seven pairs." The actual Hebrew text reads "seven seven, a male and its mate" and I believe it means "seven pairs of animals—male and female." Though many commentators have not seen here a requirement that the text means fourteen animals, it should be noted that all animals are said to have been brought onto the ark in pairs—male and female. When the text says "seven of each kind," it is always directly followed by subdividing them into pairs of male and female. Seven animals cannot be separated into even groups of male and female. Some commentators argue that the extra animal was included for the purpose of the post-flood sacrifice, but I do not believe that is tenable for the very reason that the text is clear that all animals were brought onto the ark in pairs of "a male and its mate." Secondly, in the construction for the unclean animals that were to be brought onto the ark in "twos," the text does not use the same construction and state "two twos."
8 Brown, Walt. *In the Beginning: Compelling Evidence for Creation and the Flood,* 7th ed. Phoenix, Arizona: Center for Scientific

Creation, 2001. See also his video, *God's Power & Scripture's Authority* (Phoenix, Arizona: Center for Scientific Creation, 1993.) for an animated illustration of this event.

9 Ross, Hugh, Ph.D. Audiotape: "The Flood," Part 1. Pasadena, California: Reasons to Believe, 1990.
10 Ross. Video: *The Universal Flood.*
11 Ibid.
12 Ibid.
13 See especially Corbin, B. J. *The Explorers of Ararat and the Search for Noah's Ark*. Long Beach, California: Great Commission Illustrated Books, 1999, and Bright, Richard. *The Ark, A Reality?* Guilderland, New York: Ranger Associates, Inc., 1989.
14 Recounted in personal telephone conversation (Sept 2003).
15 Cornuke, Robert and David Halbrook. *In Search of the Lost Mountains of Noah*. Nashville, Tennessee: Broadman & Holman Publishers, 2001, 59.
16 For more specifics and for interview transcripts see, Shockey, Don. *Agri-Dagh, Mount Ararat: The Painful Mountain*. Fresno, California: Pioneer Publishing Company, 1986. See also a film by Robin Simmons, *Riddle of Ararat*. Adams/Simmons, 2000.
17 Recounted to me by Richard Bright in personal conversation and in e-mail correspondence. Richard has also preserved a personal letter from Vince that reiterates his conviction that he clearly saw Noah's Ark on Mt. Ararat in eastern Turkey.

Chapter 7

THE GEOLOGY OF THE FLOOD

In considering not only the biblical account of Noah's Ark, but the nature of the geologic record of the earth's surface, it soon becomes apparent that there are really only two options: (1) the fossil record demonstrates millions of years of time and almost innumerable destructive events (most "events" sufficient to produce the type of vast fossil layers that exist in the sedimentary rocks would betoken something destructive on a scale not currently observable anywhere on earth), or (2) it is the record primarily of one worldwide, cataclysmic event (the Flood of Noah). There have been no other serious alternatives advanced by either scientists or theologians. If the biblical account of a global flood examined in the last chapter represents the true history of the world, it should be apparent to even the most casual observer that the geological features of the earth must contain evidence of that flood. This chapter, though not intended as an exhaustive look at the geological record, will nevertheless demonstrate the cataclysmic nature of the geological strata in support of a global flood.

Immanuel Velikovsky, author of several works illustrating the catastrophic nature of earth's history (*Earth in Upheaval*, *Worlds in Collision*, *Ages in Chaos*), was adept at collecting and collating data from numerous and diverse sources. Among those cited by Velikovsky was William Buckland, professor of geology at the University of Oxford (one of the foremost authorities on geology of the first half of the nineteenth century) who published a book called *Relics of the Flood*. Buckland, though he later abandoned a literal interpretation of the biblical account of the flood and became a supporter of an "ice age," had actually accumulated a wealth of geological data in support of the Noahic Flood.

Dinosaurs on the Ark

As Buckland noted, "In a cave in Kirkdale in Yorkshire, eighty feet above the valley, under a floor covering of stalagmites, he found teeth and bones of elephants, rhinoceroses, hippopotami, horses, deer, tigers ... bears, wolves, hyenas, foxes, hares, rabbits, as well as bones of ravens, pigeons, larks, snipe, and ducks. Many of the animals had died 'before the first set, or milk teeth, had been shed.'"[1]

Furthermore, "It appeared that hippopotamus and reindeer and bison lived side by side at Kirkdale; hippopotamus, reindeer, and mammoth pastured together at Brentford near London. Reindeer and grizzly bear lived with the hippopotamus at Cefn in Wales. Lemming and reindeer bones were found together with bones of the cave lion and hyena at Bleadon in Somerset. Hippopotamus, bison, and musk sheep were found together with worked flint in the gravels of the Thames Valley. The remains of reindeer lay with the bones of mammoth and rhinoceros in the cave of Breugue in France, in the same red clay, encased by the same stalagmites."[2]

Two major problems are quickly apparent: (1) the presence of what would now be considered tropical and nonmigratory animals in cold, northern climes and (2) the undecayed, unfossilized condition of the bones, which indicated a fairly "recent" event (within, certainly, thousands of years).

In many cases, the remains of plants and animals from what are now considered vastly different climates and eco-systems are found jumbled together cataclysmically. As Velikovsky puts it: "What could have brought, together or in quick succession, all these animals and plants, from the tundra of the Arctic Circle and from the jungle of the tropics, from lush oak forest and from desert, from lands of many latitudes and altitudes, from fresh-water lakes and rivers, and from salt seas of the north and the south? The shells with closed valves furnish evidence that the mollusks did not die a natural death but were buried alive.

"It would appear that this agglomeration was brought together by a moving force that rushed overland, left in its wake marine sand and deep-water creatures, swept animals and trees from the south to the north, and then, turning from the polar regions back toward the warm regions, mixed its burden of arctic plants and animals in the same sediment where it had left those from the south. Animals and plants of land and sea from various parts of the world were thrown together, one group upon another, by some elemental force [the flood!] that could not have been [merely] an overflowing river."[3]

In the 1830s, Hugh Miller carefully investigated the "Old Red Sandstone" in Scotland and concluded that it displayed "a vast stratum of water-rolled pebbles, varying in depth from a hundred feet to a hundred yards ... in a thousand different localities, ... [as well as] porphyries of vitreous fracture that cut glass as readily as flint, and masses of quartz that strike fire quite as profusely as from steel,—are yet polished and ground down into bullet-like forms... And yet it is surely difficult to conceive how the bottom of any sea should have been so violently and so equally agitated for so greatly extended a space ... and for a period so prolonged, that the entire area should have come to be covered with a stratum of rolled pebbles of almost every variety of ancient rock, fifteen stories' height in thickness."[4]

"[Also,] in the red sandstone an abundant aquatic fauna is embedded. The animals are in disturbed positions. At the period of the past when these formations were composed, 'some terrible catastrophe involved in sudden destruction the fish of an area at least a hundred miles from boundary to boundary, perhaps much more... The figures are contorted, contracted, curved; the tail in many instances is bent around to the head; the spines stick out; the fins are spread to the full, as in fish that die in convulsions... The remains, too, appear to have suffered nothing from the after-attacks of predaceous fishes; none such seem to have survived. The record is one of destruction at once widely spread and total.'"[5] As Velikovsky asks with Miller, "what agency of destruction could have accounted for 'innumerable existences of an area perhaps ten thousand square miles in extent [being] annihilated at once?'"[6] Of course, neither Velikovsky nor Miller gave credence to the biblical account of Noah's Flood, and it hardly seems necessary to state that all of their observations and questions are answered by the event of a global flood that covered even the highest mountains on earth and endured for over a year.

In fact, similar pictures are found in different places in formations both similar and dissimilar all over the globe on every continent. In northern Italy, Buckland wrote that "the circumstances under which the fossil fishes are found at Monte Bolca seem to indicate that they perished suddenly... The skeletons of these fish lie parallel to the laminae of the strata of the calcareous slate; they are always entire, and closely packed on one another... All these fishes must have died suddenly ... and have been speedily buried... From the fact that certain individuals have even preserved traces of colour upon their skin, we are certain that they were entombed before decomposition of their soft parts had taken place."[7]

These types of formations have been recorded in Scotland, Germany in the Harz Mountains, various parts of Europe, in North America in such diverse places as parts of Ohio, Michigan, Arizona, and California, in Switzerland, Italy, and numerous other places. The one thing that all these formations have in common is that they bespeak cataclysmic events the likes of which have not been recorded in modern times.

Another inexplicable phenomenon of the fossil record is the location of many fossils. For instance, whale skeletons have been found in Michigan, in Vermont, in the Montreal-Quebec area, and in great numbers in Alabama and other Gulf states. In fact, it is said, that at one time "the bones of these creatures covered the fields in such abundance and were 'so much of a nuisance on the top of the ground that the farmers piled them up to make

The Geology of the Flood

fences.'"[8] Undoubtedly, numerous sea creatures were not buried in sediments and were left floundering when the floodwaters receded from the earth.

Joseph Prestwich, professor of geology at Oxford (1874–88), studied the geology of England and came to the conclusion that at least the entire south of England as well as numerous other locales had been fairly recently submerged in water to a depth of at least a thousand feet. The fissures of limestone rocks (limestone is precipitated under water) he found were "filled with rock fragments, angular and sharp, and with bones of animals—mammoth, hippopotamus, rhinoceros, horse, polar bear, bison. The bones are 'broken into innumerable fragments. No skeleton is found entire. The separate bones, in fact, have been dispersed in the most irregular manner, and without any bearing to their relative position in the skeleton. Neither do they show wear, nor have they been gnawed by beasts of prey, though they occur with the bones of hyena, wolf, bear, and lion.'"[9]

As Prestwich observed, "If the crevices were pitfalls into which the animals fell alive, then some of the skeletons would have been preserved entire. But this is 'never the case.' 'Again, if left for a time exposed in the fissures, the bones would be variously weathered, which they are not. Nor would the mere fall have been sufficient to have caused the extensive breakage the bones have undergone ...'"[10] Other examples cited by Velikovsky demonstrate that "fissures in the rocks, not only in England and Wales, but all over western Europe, are choked with bones of animals, some of extinct races, others, though of the same age, of races still surviving... They contain remnants of mammoth, woolly rhinoceros, and other animals. These hills are often of considerable height... On the Mediterranean coast of France there are numerous clefts in the rocks crammed to overflowing with animal bones... The bones [are] all broken into fragments, but neither gnawed nor rolled. No coprolites (hardened animal feces) were found, indicating that the dead beasts had not lived in these hollows or fissures."[11]

According to Velikovsky, the Rock of Gibraltar is full of crevices that contain the broken and splintered bones of numerous animals. Examining these cracks and crevices, Prestwich discovered that "the remains of panther, lynx, caffir-cat, hyena, wolf, bear, rhinoceros, horse, wild boar, red deer, fallow deer, ibex, ox, hare, rabbit, have been found in these ossiferous [geologic deposit containing bones] fissures... A great and common danger, *such as a great flood*, alone could have driven together the animals of the plains and of the crags and caves [italics added]."[12]

Furthermore, according to Prestwich, whatever calamity destroyed and deposited all of these various animal remains together, it also impacted people since human remains and artifacts are also included among the detritus: "A human molar and some flints worked by ... man, as well as broken pieces of pottery ... were discovered among the animal bones in some of the crevices of the Rock."[13]

In fact, "on Corsica, Sardinia, and Sicily, as on the continent of Europe and the British Isles, the broken bones of animals choke the fissures in the rocks. The hills around Palermo in Sicily disclosed an 'extraordinary quantity of bones of hippopotami ...' 'No predaceous animals could have brought together and left such a collection of bones.'... 'The bones are those of animals of all ages down to the foetus, nor do they show traces of weathering or exposure.'"[14] According to Prestwich, no explanation for these phenomena would suffice other than the fairly recent submergence of at least central Europe and England, the Mediterranean islands, Corsica, Sardinia and Sicily.[15]

As Velikovsky states: "Everywhere the evidence betokens a catastrophe that occurred in not too remote times and engulfed an area of at least continental dimensions. Great avalanches of water loaded with stones were hurled on the land, shattering massifs, and searching out the fissures among the rocks, rushed through them, breaking and smashing every animal in their way."[16]

The Geology of the Flood

One other item of note that is often found in these jumbles of shattered and scattered animal bones is the remains of long-extinct animals intermixed with the remains of animals still in existence. Bones of animals from what are generally considered vastly different eras are also found intermingled. Even catastrophist Velikovsky has difficulty dealing with this "anomaly." He states, "Also bones of animals already extinct in earlier epochs were carried out of their beds and thrown into the jumble."[17] In other words, whatever event or events this betokens, according to Velikovsky's interpretation, it must have dug up the graves of long-buried and long-extinct animals and mixed them in with the freshly killed. There is no evidence for this "time-separation" in the grouping of the bones themselves, but some prejudices are not readily overcome by evidence. In fact, Velikovsky's interpretation is belied by his own findings as he often admits that these bones appear indistinguishable in age and condition.

It should probably be noted that most, if not all, of the geologists that Velikovsky quotes extensively (like Prestwich and Buckland) wrote and researched in the nineteenth century. Little of their information is popular or even known today—certainly not among current geologists. Modern geology books generally gloss over or totally ignore these dilemmas. The primary reason can be traced to the influence of uniformitarian geology (Lyell) and evolution theory (Darwin), which began to take root in the mid-1800s. Darwin, looking at the fossil record, which is a worldwide record of deaths and destruction, recognized the difficulty such observations presented. As he observed: "What, then, has exterminated so many species and whole genera? The mind at first is irresistibly hurried into the belief of some great catastrophe; but thus to destroy animals, both large and small, in Southern Patagonia, in Brazil, on the Cordillera of Peru, in North America up to Behring's [Bering's] Straits, we must shake the entire framework of the globe... It could hardly have been a change of temperature, which at about the same time destroyed the inhabitants of tropical, temperate, and arctic latitudes on both sides of the globe... Certainly, no fact in the long history of the world is so startling as the wide and repeated exterminations of its inhabitants."[18]

Darwin took his cue from Lyell, who said, "All theories are rejected which involve the assumption of sudden and violent catastrophes and revolutions of the whole earth, and its inhabitants ..."[19] That means that it is the presuppositional undergirding of uniformitarian geology that a worldwide flood could never have happened; and, therefore, visible evidences notwithstanding, the geological record is interpreted as the result of millions of years of slow processes whether such a thing is even feasible or not. But as previously cited, Dr. Derek Ager points out, "The hurricane, the flood or tsunami may do more in an hour or a day than the ordinary processes of nature have achieved in a thousand years."[20]

One question that seems to be ignored by most traditional geologists is whether there is any discernible difference between the appearance of a geologic formation that has formed slowly over vast eras of time and one that was formed rapidly by a cataclysmic event. It should be self-evident that no one has ever observed geological formations being deposited over millions of years, so most such interpretations must remain in the category of speculative fiction. On the other hand, flood deposits and cataclysms (such as the volcanic eruption of Mt. St. Helens) have been observed, and there appear to be numerous features in the earth's geological strata that clearly coincide with such events and therefore indicate rapid formation processes:

(1) Fossils. Fossils do not form except by rapid burial, either by volcanic or flood processes. Animals left on the surface decay too rapidly to be slowly buried and fossilized. Every continent of the globe bears evidence of fossil deposits. Some such fossil beds cover thousands of square miles, indicating a deposition event unlike any currently taking place—or even within recent historical memory.

The Geology of the Flood

In the fossil record are found virtually every animal currently alive on earth as well as numerous apparently extinct types (though see chapter "Dating a Dinosaur" for evidence that many animals previously thought to be extinct may still exist in small numbers—including ones now referred to as dinosaurs). We also find many "out-of-place" fossils such as human artifacts in coal beds that are buried under rock layers that are supposedly millions of years old, as well as human footprints in the same rock formation with dinosaur tracks (see chapters "Dating a Dinosaur" and "Archaeological Anomalies").

(2) Strata. We find vast layers of different types of rock in complete conformity with each other with no evidence of any erosion or soil between them. These layers were laid down either at the same time by one event, or within a very short time period. In addition, we find polystrate fossils—especially trees—in diverse places. Sometimes these trees are found upside down and stripped of all branches and roots. It is simply inconceivable that a tree was buried upside down in shallow sediment, and then endured undecayed for millions of years while layers of sediment slowly accumulated around it. Such objects virtually cry out for a cataclysmic deposition process that formed all the layers in a single event. In fact in some places trees have been discovered extending through several coal beds that are separated by sediment layers, meaning all these layers had to have been deposited very rapidly. Furthermore, some entire mountain ranges have rock layers that are folded and distorted uniformly, indicating that the entire formation was compressed while still in an unsolidified state (rocks are notoriously unbendable!).

(3) Mammoths. We have found numerous woolly mammoths frozen into the tundra of Siberia, some with the undigested remains of tropical plants still preserved in their system. Their condition indicates immediate burial in a cataclysmic process that plunged them to a temperature of at least -150 degrees Fahrenheit within minutes. No layers of sedimentary fossil deposits have yet

been discovered under the frozen mammoths, indicating they were buried immediately prior to the rising of the floodwaters in the Noahic Flood (see chapter "The Myth of the Ice Age").

Velikovsky, who catalogued and brought together much evidence of cataclysmic destruction from all over the globe, had numerous explanations for them, including asteroids crashing into the earth and Mars passing too close for comfort. Without a historical document or recording to guide the geologist, any number of explanations are possible—though none of them are either testable or demonstrable. There is, however, a historical record of an event that not only fits the observable record, but explains much that is otherwise inexplicable. It happens to be the biblical record of the Flood of Noah (see chapter "The Biblical Account of the Flood").

There have been three basic alternative interpretations of this account offered in Christian writings: (1) this account is of no historical significance—it has only some spiritual truths to convey (of course nothing in the text indicates this even by implication), (2) the account is historical, but it was a "local flood" that was thought by Noah and his family (or later authors) to include the whole world, or (3) it is a historical account of a cataclysmic flood of global dimensions.

Of the first alternative, I would only ask, what criteria are used to reject its roots in history? If it is the implausible or even miraculous nature of the event, then you must be willing to reject the Incarnation and the Resurrection as well. As far as the second alternative goes, if this was a local flood that covered only the Mesopotamian valley, I think an honest observer would find that it is impossible to glean that from the text alone. In verses 21–23 it repeats three different times in three different ways that the flood was intended to wipe out every creature on the sphere of the earth that had "the breath of life in its nostrils." Some have argued that this story is told from man's perspective and therefore it was only the portion of the world that men knew and could see that was destroyed. But if you also look at 6:13, you will see that it is God who states, "I am surely going to destroy [people] and the earth." And, once again, the only reason for trying to interpret it in any other way is in order to maintain the fossil record as the record of hundreds of millions of years of time and uniformitarian geology. If the fossil record is a record of, primarily, one great flood, then all evidence for millions of years of time is obliterated. There is, indeed, much at stake in locating the remains of Noah's Ark on the top of Mt. Ararat.

In summary, there is a wealth of geologic evidence that clearly supports the worldwide and cataclysmic nature of Noah's Flood. Every continent is covered by sedimentary deposits over a mile thick; fossils and animal bones that show evidence of massive waterborne destruction litter every country on earth; marine fossils are found even at the upper levels of most mountain chains on earth; layer after layer of sedimentary deposits and coal and petroleum beds betray evidence of the cataclysmic nature of their deposition, and many layers have been found to contain human artifacts or "anomalous" fossils (see chapter "Archaeological Anomalies"). Indeed, the different layers of the fossil beds attest, not to different epochs, but to different environmental habitats as well as the tremendous sifting capacity of water. The animals we call "dinosaurs" are a part of that same fossil record and, for the most part, are animals that eventually became extinct in changed conditions following the flood (though, see chapter "Dating a Dinosaur" for evidence that some of the major dinosaur kinds may still exist in small numbers in isolated populations).

Some commentators have argued against this by saying it would have been cruel of God to spare dinosaurs from the flood only to allow them to eventually die out afterward. But the same thing is happening today with many animals that are not categorized as "dinosaurs." So it would have to be considered equally cruel for God to create any species that would die out at any time. Some have apparently forgotten—or never considered—that the earth and the animal kingdom are suffering the effects of the curse that was the result of man's sin.

Furthermore, the dimensions given for the ark indicate that it was enormous. Not only did it have tremendous carrying capacity, but tests have shown it would have been virtually unsinkable (In World War II the Navy produced a destroyer with the same proportions as the ark and found it to be their most stable ship ever.). In fact, the ark may have been the world's largest ship ever, up until the ocean liners and aircraft carriers of the twentieth century. The most thorough study to date has demonstrated conclusively that it clearly had the capacity to carry not only the requisite animals, but the requisite food, water, supplies, and people as well. I believe that, when the ark is again discovered, mankind will be astonished at the construction of the ark: the systems in place for caring for the animals, for supplying food and water, as well as the cleaning and ventilation. The Bible makes it clear that the flood was intended to destroy the entire face of the earth under the heavens—including all mankind and animals with the breath of life in them, except for those spared on the ark, which then came to rest on the mountains of Ararat. The only reason for rejecting such an interpretation is, once again, the advent of evolution theory and naturalism, which demanded long eras of time for dinosaurs to evolve and die out (from what?!).

Therefore, the key to re-interpreting the geological record of the earth, as well as discerning the nature of creation, the scriptural record, and even, in some respects, the heart of the Creator Himself, lies buried in the ice somewhere in the upper reaches of Mt. Ararat; preserved by the Creator for such a time as this.

Endnotes

1 Buckland, William. *Reliquiae Diluvianae*. Cited in Velikovsky, Immanuel. *Earth in Upheaval*. Garden City, New York: Doubleday & Company, Inc., 1955, 16.

The Geology of the Flood

2 Dawkins, W. B. *Proceedings of the Geological Society.* 1896, 190, and Cave-hunting. 1874, 416, and Geikie, James. *Prehistoric Europe.* 1881, 137, cited in Velikovsky, 16–17.
3 Velikovsky, 58.
4 Miller, Hugh. *The Old Red Sandstone.* Boston, 1865, 217–218, cited in Velikovsky, 19.
5 Miller, 222, cited in Velikovsky, 19–20.
6 Miller, 223, cited in Velikovsky, 20.
7 Buckland, William. *Geology and Mineralogy.* Philadelphia, 1837, 101, cited in Velikovsky, 21.
8 Price, George McCready. *Common-sense Geology.* 1946, 204–205, cited in Velikovsky, 48.
9 Prestwich, Joseph. *On Certain Phenomena to the Close of the Last Geological Period and on Their Bearing upon the Tradition of the Flood.* London: Macmillan and Co., 1895, 25–26, cited in Velikovsky, 50–51.
10 Prestwich, 30, cited in Velikovsky, 51.
11 Velikovsky, 51–52.
12 Prestwich, 46, cited in Velikovsky, 53.
13 Prestwich, 48, cited in Velikovsky, 53.
14 Prestwich, 50–51, cited in Velikovsky, 53–54.
15 Velikovsky, 54.
16 Ibid., 55.
17 Ibid., 58.
18 Darwin, Charles. *Journal of Researches into the Natural History and Geology of the Countries Visited During the Voyage of H.M.S. Beagle Round the World.* (January 9, 1834), cited in Velikovsky, 33.
19 Lyell, Sir Charles. *Principles of Geology,* 12th ed. 1875, 318, cited in Velikovsky, 27.
20 Ager, Derek. *The Nature of the Stratigraphical Record.* New York: John Wiley & Sons, 1993, 80.

Chapter 8

THE FOSSIL RECORD

In general, a fossil is a plant, insect, or animal part that has been permineralized—basically, turned into "stone" by a process of mineralization. If it is the fossil of an animal skeleton, the bones are no longer bones, they are rock made of various minerals. Usually, geology texts state that to become fossilized, a plant or an animal must have had hard parts (bones, branches, etc.) and become buried quickly (note: this is not always the case since even jellyfish have been found fossilized—these would generally evaporate in a very short period). Rapid burial is generally cited as essential in order to prevent decay (note: usually in waterborne sediment), and the mineralization process is most often said to have taken place over a very long period of time. The fossil record, whatever else it may be, is always and only a record of death and destruction—usually on a massive scale. It is interesting, to say the least, that the general public perception seems to be that it is the fossil record that provides the essential evidence of evolution, when in reality the fossil record represents the demise of any remnant of hope for the theory.

One of the main misconceptions about the fossil record is that fossils have somehow been accurately dated, thus establishing a record of the earth's crust at hundreds of millions of years. But the question is, how is it possible to date a fossil? One popular misconception is that carbon-14 dating methods are somehow used. Besides the unsupported presuppositional problems inherent in C-14 dating of any kind, since this process measures radioactive carbon present in a once-living object, it is useless in dating a true fossil: there are no carbon atoms present in a fossil since they have been replaced by other minerals.

In addition, since any measurable amount of carbon-14 would be gone after approximately fifty thousand years, it would be useless as a method to date any object that is actually millions of years old.[1] Furthermore, sedimentary rocks, in which virtually all fossils are found, do not lend themselves to any other type of radioactive dating methods because they generally do not contain other radioactive materials. Consequently, the dating of dinosaur fossils (see chapter "Dating a Dinosaur"), or any other fossils, is not just suspect, it is fiction.

The truth is that most fossils are dated based on the idea of evolution. If we find, for instance, a fossilized creature that theoretically represents a particular stage of evolution, and we know how long ago that organism became extinct (how do we "know" without having recorded or observed the process?), then we can generally date any layer in which we find that particular fossil. Of course, all this is conjecture based on a belief in evolution and not on any objective facts. For example, on a televised edition of *Nightline* (with Ted Koppel), entitled "Fish Fingers," two scientists reported finding the fossilized remains of what they described as a "fish fin" that had an internal bony structure that, to them, was reminiscent of human fingers. They postulated that these fish "fingers" were used by the fish to crawl along the bottom of the sea (who observed such a thing?) and when the fish one day crawled out on land, the fins eventually became useful as hands (what on earth did they do in the meantime while they waited for lungs to develop?). The scientists assigned a date to the fossil of 370 million years, based primarily on their idea that this was the time period such a process must have taken place (presumably). They even managed to draw the fish to which the fin was once attached. It was very large; on the scale of six feet long, with very large jaws. However, at least a couple of "minor details" were omitted or ignored in the making of this show: (1) Only the fin fossil was ever shown, and if any of the rest of the fossil existed it was never depicted. So the artist who manufactured the drawing could have made it appear however he/she chose, and we would never be the wiser. (2) The fossil was found, along with numerous other fossils, *in the mountains* of central Pennsylvania. No one bothered to explain how or when Pennsylvania was last covered by a flood or an ocean (!), or, better yet, how a fish happened to be buried in sediment while crawling in the mountains of Pennsylvania. The idea that this fossil was 370 million years old depends absolutely on the theory of evolution being true, and on the scientists having interpreted every aspect of it absolutely correctly. So how do we know that the interpretation or "placement" of a fossil is correct? The truth is, unless we can examine the organism itself, it is simply impossible. Any interpretation of a fossilized animal that is currently unknown in the living record remains highly speculative at best.

A number of years ago, another unusual fish fossil was discovered. This one was named "coelacanth" and was said to have fish "legs." On that basis, at the time, it was said to date from the Cretaceous period about one hundred million years ago. Because it appeared to have appendages that were neither fins nor legs, it was claimed that the appendages were fins on the way to becoming legs. Coelacanth was then said to represent the missing link between fish and amphibians, making it the only known "link" to fill that tremendous gap. Of course, such claims made on the basis of a speculative interpretation of fossilized external features are highly subjective at best. The fossil can neither be dissected in order to examine internal organs, nor observed in action. Nevertheless, *The New World Dictionary* (1976) defined the coelacanth as "any of an order of primitive marine fishes, possibly ancestors to land animals ..."[2] Bear in mind, that "definition" occurred thirty-eight years *after* a fisherman caught a live one that was still swimming in the ocean! Since that time, numerous living coelacanth have been found, and they have now been physically examined in life and death. In fact, it is now known that virtually everything scientists once conjectured about them, based

The Fossil Record

solely on the fossil record and the idea of evolution, was absolutely wrong. The coelacanth does not have partial or developing lungs (which had been postulated in order to explain how they were able to use their "legs" to crawl out on land)! They are only fish. Not amfishians! They are still around and still swimming, not crawling on land or in the sea. They are not only not a missing link, they are not even missing.

Top: current coelacanth. Bottom: "100-million-year-old" fossil.
[Photograph from Nelson, Byron. *After Its Kind*. Minneapolis, Minnesota: Bethany Fellowhip, Inc., 1967, p. 56.]

Furthermore, it means that since they did not become extinct seventy to one hundred million years ago, the fossils that have been found could have been formed only a few thousand years ago—or less. But no one has gone back and changed or eliminated all the other estimated ages of the fossil layers that were dated on the basis of understanding coelacanth as a long-extinct missing link. In fact, my 1996 dictionary defines coelacanth as "a heavy, hollow-spined fish ... of deep S. African coastal seas, that crawls on the sea bottom with lobed, limb-like fins: a living fossil ... considered forerunners of the land vertebrates." Note that even though it is admittedly a "fish," it is claimed that it "crawls along the bottom" with its leglike fins. Is this even true? Who watched it? Has it been filmed, or is this "definition" the result of evolutionary beliefs and wishful thinking that continue sixty years after discovery should have eliminated all questions? The truth is that the coelacanth has now been observed and filmed in its natural habitat and it is utterly false that it uses its limbs to crawl! So, how and why was that statement even a part of the "definition" when there was not only no evidence for such a thing, but it was falsifiable? And, as it turns out, false! It is because the coelacanth, a key index fossil that was once used to bolster the idea of evolution, must continue to occupy that ground or the whole dating method is called into question. Over sixty years after its discovery it was still being defined in a way that made it seem like a possible "missing link." Indeed, it is called a "living fossil," but if it still exists today, unchanged by any evolutionary process, how can it in any way be claimed to be "primitive"? How can the organism that supposedly evolved into amphibians exist side-by-side with all current amphibians and fish? Something is indeed fishy here! (Something am-phishy here?)

One of the most important issues that needs to be addressed is how and why the fossil record is so important to the theory of evolution in the first place. First of all, it should be clear that the living record is, itself, an insoluble problem for the theory of evolution. The very existence of species mitigates against evolution; for, if animals gradually and incrementally changed, we should find an infinite "blend" of one animal into another. There should be no possibility of distinct and unique species even existing. For instance, if human beings emerged over the course of millions of years from some type of "apelike" ancestor, where are those creatures? Each "missing link" should have been more advanced, more intelligent, and more capable than its predecessor. After all, that is precisely how the theory accounts for their survival. However, the less intelligent, less capable, less advanced gorillas, apes, orangutans, chimpanzees, and so on, are all still around. How is it possible that all the links, and only the links, died out everywhere?

Darwin himself recognized the extreme difficulty in explaining this away (though try he did): "Long before the reader has arrived at this point of my work, a crowd of difficulties will have occurred to him. Some of them are so serious that to this day I can hardly reflect on them without in some way being staggered ... [Question: if there were so many "staggering" difficulties with the theory, why did he work so hard to insist on its plausibility in spite of the overwhelming obstacles?] First, why if species have descended from others by fine gradations, do we not everywhere see innumerable transitional forms? Why is not all nature in confusion, instead of the species being, as we see them, well defined?"[3] Darwin's question remains scientifically unanswered and apparently, unanswerable.

Ernst Mayr, evolutionist and professor at Harvard, did experiments with fruit flies in order to demonstrate "evolution." (His thesis: since fruit flies reproduce so rapidly—a new generation every eleven days—if evolution were possible, it should be demonstrable in a laboratory. A hundred years of fruit fly populations should be the equivalent of more than a hundred thousand years of human populations.) Mayr selected one feature to change—"evolve": the normal fruit fly has thirty-six hairlike bristles. After bombarding them with radiation, the flies with more than the normal number of bristles were hand-selected out (notice that this is no longer "evolution" because the process is being artificially directed by "intelligence"). Researchers discovered that there was apparently an upper limit on the amount of genetic variation that could take place. When the genetic plateau of fifty-six bristles was reached, no more increase could be achieved without killing off the fruit flies. This fact alone would necessarily preclude evolution. When that population of fruit flies was left alone, within a few years they had returned to normal. Researchers also attempted to decrease the number of bristles by selecting out fruit flies with less than the normal number. Yet once again they found limitations and genetic parameters beyond which they could not go (they achieved a minimum of twenty-five bristles).

Essentially, they discovered that the genetic makeup of the fruit fly was engineered to prevent evolution from happening! After more than a hundred years of genetic research with fruit flies, scientists have nothing to show for their research but fruit flies. Fruit flies remain fruit flies remain fruit flies.[4]

The only possible salvation for the theory must then lie in the fossil record. Many laymen are under the mistaken impression that evolution theory began as an attempt to explain what was found in the fossil record. In fact, nothing could be further from the truth. Indeed, many people are surprised to discover that Darwin himself felt that the fossil record as it existed in his time, was the best evidence against evolution. As he stated: "Just in proportion as this process of extermination has acted on an enormous scale, so must the number of intermediate varieties, which must have formerly existed, be truly enormous. Why then is not every geological formation and every stratum full of such intermediate links? Geology assuredly does not reveal any such finely-graduated organic chain; and this, perhaps, is the most obvious and serious objection which can be urged against the theory. The explanation lies, as

I believe, in the extreme imperfection of the geological record."[5] In other words, realize that Darwin is advancing his theory in spite of his own understanding that the existing evidence—living and dead—was against it. And the only hope for his theory, as he saw it, was that the fossil record had not yet been adequately examined. Yet, as David Raup (curator of geology, Field Museum of Natural History, Chicago) admitted, "Darwin's theory of natural selection has always been closely linked to evidence from fossils, and probably most people assume that fossils provide a very important part of the general statement that is made in favor of Darwinian interpretations of the history of life. Unfortunately, this is not ... true."[6]

It is somewhat astonishing that in the modern era, many laymen seem to be under the impression that Darwin's theories on evolution were gleaned from, or at least supported by, the evidences he accrued. Yet even his most ardent supporters will generally admit that Darwin produced virtually no concrete evidence to support any of his primary claims. The fossil record, as he himself admitted, not only showed no evidence of transition from one type of creature to another, but any current species found duplicated in the fossil record remained clearly identifiable as the same unchanged creature. In addition, he failed to observe or identify a single evolutionary transformation among living species (such as the production of wings where there previously were none). In fact, Darwin did not even attempt to address the main question of how life came to be in the first place. If life itself cannot happen by random chance, then the question of how it might change over time is rendered completely meaningless. It is tantamount to addressing how mistakes on the assembly line might, given enough time, accidentally change a passenger car into a pickup, while ignoring the more crucial question of how the automobile and the assembly line itself came into existence in the first place. According to Darwin, then, the problem lay in simply having not uncovered enough of the fossil evidence. He reasoned that, once we began to look more thoroughly, we would soon find the millions of necessary transitional links. Unfortunately for the theory, after 150 years of desperate searching, missing links are still missing. Even worse, some of the former so-called transition links have come back to life, and are no longer either links or missing.

The very fact that new explanations of evolution have been advanced—such as Goldschmidt's "Hopeful Monster Mechanism," and punctuated equilibrium—should make it obvious that the issue of missing links has not been resolved. The so-called "Hopeful Monster Mechanism" was proposed by geneticist Richard Goldschmidt precisely to explain evolution while allowing for the lack of fossil evidence of same. Basically, what he proposed was that evolution happened in tremendous leaps of genetic change in one generation without leaving any evidence of having taken place. In other words, he said that, in essence, one day a reptile laid an egg, and a bird was hatched. Besides the fact that this would require truly miraculous genetic changes, all working together, it is not very confidence-inspiring on the practical level. If a bird were truly hatched by a snake, who would feed it and care for it into adulthood? The mother snake would be far more likely to eat the bizarre offspring than feed and care for it. And if she somehow overcame her natural instincts and tried to feed the bird, what would she feed it? Do baby snakes maintain the same diet as baby birds? Hardly! In reality, mother snakes do not even feed or care for their offspring at all. And even if these major obstacles were somehow overcome, a far greater hurdle for evolution to overcome would be reproduction. Most evolutionary texts make no attempt to explain the existence of coordinating reproductive organs. One bird alone would soon die out even if it did manage to survive into adulthood. And one bird cannot mate with just any other bird—it must be a bird of the same kind but opposite sex with

The Fossil Record

perfectly coordinating reproductive systems. Unless a comparable female bird with coinciding reproductive systems and organs was born simultaneously, it would be the end of the miracle. It may sound ridiculous, and it did not achieve a very wide following, but at least it was an open admission of the total lack of fossil evidence for gradual change over time.

Nowadays, Goldschmidt's "Hopeful Monster Mechanism" has been replaced by a theory called "punctuated equilibrium."

This theory attempts to make the lack of transitional links more intellectually palatable by postulating that evolution took place rapidly in a small, isolated portion of the population, again leaving little or no evidence. The improved variety of species then moved into the neighborhood of the original species, which then caused the original variety to rapidly die out. But the question then becomes, if punctuated equilibrium occurs without leaving any evidence behind, how can such a theory ever be tested or falsified? Because, if life evolved, it is clear that it managed to do so without leaving the necessary transitional forms in the fossil record or in the living record. The very fact that theories like Goldschmidt's "Hopeful Monster Mechanism" and "punctuated equilibrium" have been advanced at all should demonstrate that evolution, in contradiction to Darwin's entire thesis, cannot have happened gradually after all.

As Michael Denton states, "The absence of intermediates, although damaging [to Darwin's theory], was not fatal in 1860, for it was reasonable to hope that many would eventually be found as geological activities increased... By stressing the very small fraction of all potentially fossil-bearing strata examined in his time, Darwin was able to blunt the criticism of his opponents who found the absence of connecting links irreconcilable with organic evolution... [However], virtually all the new fossil species discovered since Darwin's time have either been closely related to known forms or ... strange unique types of unknown affinity... While the rocks have continually yielded new and exciting and even bizarre forms of life, what they have never yielded is any of Darwin's myriads of transitional forms."[7]

Ironically, Darwin himself admitted: "By the theory of natural selection all living species have been connected with the parent-species of each genus, by differences not greater than we see between the natural and domestic varieties of the same species as the present day; and these parent-species, now extinct, have in their turn been similarly connected with more ancient forms; and so on backwards, always converging to the common ancestor of each class. So that the number of intermediate and transitional links, between all living and extinct species, must have been inconceivably great. But *assuredly, if this theory be true, such have lived upon the earth* [italics added]."[8]

That means that Darwin himself left scientists with the means of falsifying his own theory. If a more thorough examination of the fossil record did not yield the necessary myriad links, then by Darwin's own words, his theory would be falsified. According to Denton, over ninety-nine percent of the fossil record has been uncovered since Darwin; and this is what the evolutionist literature says about it: "The known fossil record fails to document a single example of phyletic (gradual) evolution accomplishing a major morphological transition and hence offers no evidence that the gradualistic model can be valid."[9] Dr. Colin Patterson, senior paleontologist at the British Museum of Natural History, had this to say about the lack of illustrations of transitional forms in his book on evolution: "If I knew of any, fossil or living, I would certainly have included them... I will lay it on the line—there is not one such fossil for which one could make a watertight argument."[10]

The gaps, it appears, will not go away and they are not trivial. Filling the gaps, in reality, should have been the only hope for the theory of evolution if it were indeed just another scientific theory. As Denton states, "The gaps which separate species: dog/fox, rat/mouse etc are utterly trivial compared with, say, that between a primitive [sic] terrestrial mammal and a whale or a primitive [sic] terrestrial reptile and an Ichthyosaur; and even these relatively major discontinuities are trivial alongside those which divide major phyla such as mollusks and arthropods. Such major discontinuities simply could not, unless we are to believe in miracles, have been crossed in geologically short periods of time through one or two transitional species occupying restricted geographical areas... To suggest that the hundreds, thousands or possibly even millions of transitional species which must have existed in the interval between vastly dissimilar types were all unsuccessful species occupying isolated areas and having very small population numbers is verging on the incredible!"[11]

Indeed, it means we are insisting on believing a theory for which the evidence is not only nonexistent, but actually contrary to the predictions of the theory. The following table is taken from *Evolution: A Theory in Crisis* and it demonstrates the adequacy of the known fossil record:

Number of living orders of terrestrial vertebrates: 43

Number of living orders of terrestrial vertebrates found as fossils: 42

Percentage fossilized: 97.7%

Number of living families of terrestrial vertebrates: 329

Number of living families of terrestrial vertebrates found as fossils: 261

Percentage fossilized: 79.1%

Number of living families of terrestrial vertebrates excluding birds: 178

Number of living families of terrestrial vertebrates found as fossils excluding birds: 156

Percentage fossilized: 87.8%

8.5: The Adequacy of the Fossil Record. The table shows the percentage of living orders and living families that have been recovered as fossils.[12]

Note that, of the 329 living families of terrestrial vertebrates, 261 (almost eighty percent) have been found as fossils, and when birds, which are poorly fossilized (generally because they can fly above the conditions that would ensnare most animals), are excluded, the percentage rises to eighty-eight percent! And note also that ninety-five percent of the world's fossils are marine invertebrates. Only about one in ten thousand fossils is a vertebrate, and most of those are fish!

Nonetheless, major news publications such as *Time*, *Newsweek*, and *National Geographic* continue to publish persuasive articles touting the evidence for human evolution from apelike ancestors. If the links are truly missing, why do such stories persist? Remember, the early evolutionists, especially Darwin, did not begin with the fossil record; but they were anxious to find evidence within the fossil record to support the theory since there is clearly nothing in the living record to do so. Indeed, it is essential in examining this issue to remember two very important things: First, there are no known living specimens that clearly demonstrate an evolutionary process (that fact alone should be sufficient evidence to discount the entire theory). So any conclusions based on the fossil record alone can never be tested adequately by any scientific method. Second, people who have been looking for evidence to support the theory of evolution have generally begun with the presupposition that evolution did happen. They have searched with desperation to find anything that can be inserted into the theory with which they began. That is not the hallmark of a scientific investigation; it is a desperate search of faith.

One of the first men to find evidence to support Darwin was a young Dutch doctor named Eugene Dubois. In all such cases, it is crucial to realize that Dubois was determined to find evidence of the missing link. He found

The Fossil Record

a skullcap that was "ape-like," two humanlike teeth, and a human femur (thighbone). That is all! The bones were certainly not a complete skeleton. They were not attached to each other or even to other bones (see illustration below). There was no evidence of any civilization of these "creatures." The truth is that they were not even found in the same location. The skullcap and the legbone were found some fifty feet apart. Nevertheless, for many scientists, even to this day, this became "Java man," missing link and evidentiary proof of Darwin's theory. However, it is also crucial to understand that virtually no such find receives universal recognition, nor even general agreement, as to the era it stems from, or the kind of evolutionary link it represents. At the time of Dubois's public claims, Rudolph Virchow thought that the deep suture in the skull was just like that of an ape: "In my opinion," he stated, "this creature was an animal, a giant gibbon, in fact. The thigh bone has not the slightest connection with the skull."[13]

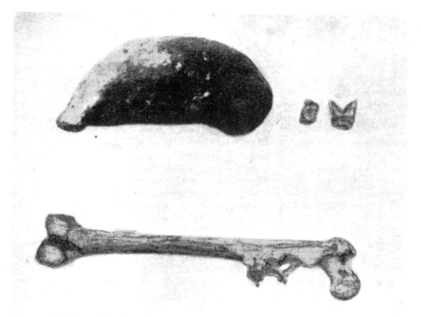

The real "Java Man" as he was originally discovered—the skull cap and femur were not even found together. [Photograph from Nelson, p. 129.]

G. K. Chesterton, Christian apologist and prolific author, dryly noted that scientists would speak of "Java Man" (Pithecanthropus) in the same vein as historical figures and that detailed drawings would depict every hair of his head. And yet "no uninformed person, looking at its carefully lined face, would imagine for a moment that this was the portrait of a thigh bone, of a few teeth, and fragment of a cranium."[14] In addition, Dubois, years later, confessed that along with the human leg bone, he had also found two other skulls, which he had not previously revealed.[15] Most scientists now concede that these skulls were fully human. Unfortunately for Dubois, they had been found in the same layer and area as his other finds, and they did not support his conclusions about evolution. So he kept them hidden for thirty years. Though there is almost never scientific consensus on such a find, it appears that most commentators now feel that the original "Java Man" skull was most likely the remains of some kind of ape.

In 1911, scientists, again looking for evidence to try and prove evolution (a very dangerous and unscientific position to be in), found some skull and jaw pieces and managed to put them together to make Piltdown man—another so-called missing link. (See following artist's rendition of Piltdown Man's appearance.) Since they did not have an entire skull, they decided how large a skull the proper missing link should have, made a plaster cast, and fit the skull pieces into it. For over forty years, the Piltdown skull was presented to the world as proof of evolution. Again, there was no civilization of these creatures, no complete skeleton, just a few fragments. But it was the best evidence available and it was heralded by many of the world's best scientific minds as definitive evidence for Darwin's theories. As the headlines of the *New York Times* proclaimed on December 22, 1912, "Darwin Theory Proved True; English Scientists Say the Skull Found in Sussex Establishes Human Descent from Apes." The key words that should be noted are "proved" and "establishes." In fact, it is reported that in forty years, over five

The Fossil Record

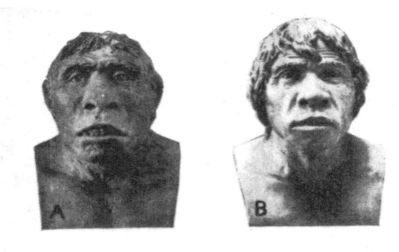

Artists' renderings of various "evolutionary links:" (A) Java Man, (B) Piltdown Man, (C) Heidelberg Man, (D) Neanderthal Man. [Ibid., p. 127.]

hundred doctoral dissertations were done on the basis of Piltdown Man. It was not until 1953 that it was finally revealed that "Piltdown Man" was actually made up of the jawbone of an orangutan stained to look old and, with teeth that had been filed flat to look more humanlike, fit together with a few human skull pieces. To this day no one knows for sure who perpetrated the fraud, but one thing is abundantly clear, it does not take much to fool those who already believe.[16]

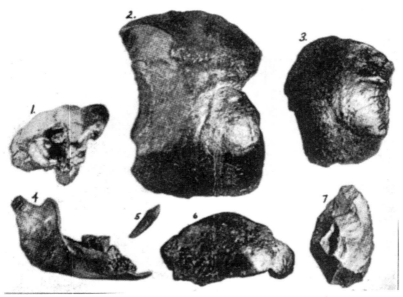

"Piltdown Man" as he originally appeared. [Ibid., p. 131.]

Another example that most evolutionists would rather not be reminded of is that of Nebraska Man, or his more erudite-sounding scientific name, Hesperopithecus. During the infamous Scopes "Monkey Trial," William Jennings Bryan, a politician and former presidential candidate from Nebraska, fought vigorously against teaching school-children the theory of evolution. Evolutionist H. F. Osborn, head of the American Museum of Natural History, ridiculed Bryan in the press by exalting over the "evidence" of Nebraska Man. Evolution was not just a pleasing theory, it was supported by the best evidence science could offer. Nebraska Man had been constructed primarily on the basis of one tooth found in so-called Pliocene deposits in the state of Nebraska. Since this tooth was deemed "humanlike," but dated to be at least one million years old, it became "Nebraska Man," or Hesperopithecus—evidence supreme of human evolution from apelike ancestors. As Osborn chortled, "The earth spoke to Bryan from his own state of Nebraska. The Hesperopithecus tooth is like the still, small voice [an apparent mocking reference to the "still, small voice"—or "gentle whisper" in NIV—of God that spoke to Elijah in I Kings 19:12]. Its sound is by no means easy to hear. But this little tooth speaks volumes of truth, in that it affords evidence of man's descent from apes."[17] Bryan had publicly derided such claims as imaginative fiction at best. As he said to the West Virginia state legislature in 1923, "If they find a stray tooth in a gravel pit, they hold a conclave and fashion a creature such as they suppose the possessor of the tooth to have been, and then they shout derisively at Moses."[18]

Apparently, Bryan had a prophetic gift as well as oratory skills: eventually the rest of the jaw and more teeth of "Nebraska Man" were discovered. They turned out to belong to an extinct pig (see illustration of artist's rendition of Nebraska Man). Bryan was absolutely right, and the "evidences" that supposedly had convinced evolutionists of the veracity of their claims were fiction and fairy tales. The "missing link" is missing from the fossil record as well as the living record precisely because it never existed. In fact, it exists only in the mind of the evolutionist. It is interesting to note that Edward Larson, who wrote a book, *Summer for the Gods*, which purported to explain the actual history of the Scopes Monkey Trial, mentions the rhetoric on both sides of the "Nebraska Man" issue in the early 1920s. He leaves the reader with the distinct impression that the fossil find was legitimate evidence of evolution that was ridiculed by creationists for no apparent reason. Nowhere, even in the extensive footnotes, does he mention that "Nebraska Man" was, in reality, *a pig*.[19]

It should be of more than passing interest that three of the main evidences presented at the Tennessee "Monkey Trial" in 1925 to bolster the case for teaching evolution in schools were the "Java Man" fossils found by Dubois, the Hesperopithecus or "Nebraska Man" fossil, and the "Piltdown Man" skull. As Francis Hitching points out, it should

"Nebraska Man" illustration. This drawing was based on the evidence of a single tooth, later discovered to be the tooth of a pig. [Fig. 14, *In the Beginning: Compelling Evidence for Creation and the Flood, 8th ed.*, 2008, by Walt Brown, p. 13, from G. Elliot Smith, *Illustrated London News*, June 24, 1922, p. 944.]

be seen as the height of irony, that "the trial that became a turning point in US educational history, not to be significantly challenged for the next half-century, was steered towards its verdict by a pig tooth [Nebraska Man], two dubious fossils subsequently repudiated by their finder [Java Man and Dubois], and an outright fake whose perpetrator is still not known [Piltdown Man]."[20]

It is almost mind-numbing to speculate how many "scientists" have earned their degrees on the basis of wishful thinking rather than evidence. Evolutionists have tried to downplay such misinterpretations of the evidence by pointing out that the fraud or error was eventually discovered, and that proves their contention that science is "self-correcting." However, one must wonder where that process was for forty years and over five hundred doctoral dissertations in the case of Piltdown Man. Certainly, if nothing else, such a case should serve to remind us that any such fossil find ought to be treated with a healthy dose of skepticism until the process of "science" has had at least forty or fifty years and a few hundred meaningless doctoral dissertations under its collective belt. One must legitimately question why none of those five hundred doctoral degrees granted on the basis of fraud and error were rescinded when the hoax was revealed. Sadly, the truth is that the mistaken experts, with nary a hitch in their stride, go right on teaching in the universities and making up new theories to fit the evidence into.

Neanderthal man is another commonly cited evidence for evolution. One major problem for the theory of evolution to overcome is that, based on the skull and skeletal remains that have been found, Neanderthal Man had an average brain size over two hundred cubic centimeters larger than today's average. If it were not for the theory of evolution that is so desperate for evidence, there would be no question that Neanderthal man was completely human.

One of the first experts to examine skeletal evidence, Rudolph Virchow (commonly referred to as the father of modern pathology), decided they belonged to a modern Homo sapien "who had suffered from rickets in childhood and arthritis in old age."[21] More recent studies seem to indicate that this may not be the case, but a very thorough study by an orthodontist named Jack Cuozzo has demonstrated conclusively that published accounts of the so-called Neanderthal skulls have not only been mistaken in their reconstructions, but intentionally misleading.[22] In fact, one of the skulls he studied and x-rayed appeared to have a bullet hole through the skull. Whether or not the wound could be proven to be a bullet hole is almost secondary to the fact that textbook photographs of the skull have been intentionally doctored to eliminate

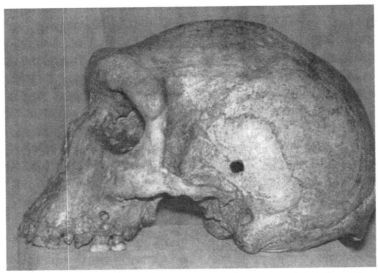

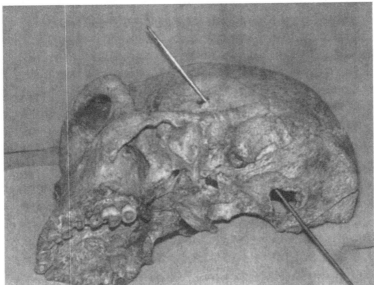

Photos showing apparent bullet hole through Neanderthal, "Broken Hill Man." [Photograph from Cuozzo, Jack. *Buried Alive: The Startling Truth About Neanderthal Man*. Greenforest, Arkansas: Master Books, 1998, p. 70.]

The Fossil Record

the hole. It certainly seems unlikely that such doctoring has been done for simply cosmetic reasons.

Cuozzo has also demonstrated that the Neanderthal skulls demonstrate men with larger, thicker, and more slowly developing skulls. Such a reconstruction would be completely consistent with the biblical account of post-flood men that lived for several centuries, and also consistent with a process of human development that is exactly the opposite of evolution. In any case, it should be extremely difficult for the evolutionist to explain how our evolutionary ancestor had a much larger brain than ours, especially considering that it is brain size that is used to show and "prove" the progression of evolution. Most often, this inconvenience to the theory is simply not mentioned. Interestingly, Cuozzo has also studied more recent skeletons, as well as medical records, to show that such "antievolution" rates are still occurring. One coffin from the mid 1700s contained the remains of a thirteen-year-old child (the age was written with brass tacks on the coffin). Dental studies of the teeth would have concluded the child was only nine or ten at most. Cuozzo demonstrates that growth rates and maturations rates are still speeding up, consistent with overall deterioration and decay processes that are the antithesis of evolution theory.[23]

A final example of evolutionary misrepresentation or misinterpretation of the evidence is a fossil that has been commonly named "Lucy." Lucy, a so-called Australopithecine, still occupies a place of prominence in most evolutionist literature. Her bones were unearthed in 1974 by Donald Johanson in Ethiopia. The remains consist of about forty percent of the skeleton of a small, female, apelike creature estimated to be about three feet tall and possessing a very small brain. Johanson declared "Lucy" to be an upright-walking, three-and-a-half-million-year-old predecessor to humans. Once again, it must be noted that not even all evolutionists are in agreement as to what kind of creature Lucy represents. Johanson claims that Lucy had a small, but essentially human body. Yet Lucy's

The actual skeletal remains of "Lucy." Note the total lack of any hand or foot bones. In addition, the "missing link" skull recreated by artists is virtually nonexistent. [Used by permission from the Houston Museum of Natural Science.]

skull was incomplete, and even Johanson conceded that a reconstruction "looked very much like a small female gorilla."[24] In fact, numerous dissenting scientists (Susman, Stern, Oxnard, among others) have challenged even the view that Lucy walked upright like a human. Many have demonstrated that the arm, shoulder, and body are all essentially apelike.[25] Perhaps, if it were not for the knee joint, even most evolutionists would concede that this creature would appear no different from current monkeys. Yet, in a 1986 lecture, Johanson admitted under questioning that the knee-joint of a supposedly similar creature that helped give Lucy her elevated status as prehuman missing link was actually found in strata two hundred feet lower than the rest of Lucy's skeleton and two to three kilometers away! As one observer dryly noted, perhaps it would be "ridiculing the case too harshly to ask if this creature lost one leg, then limped several kilometres before expiring, while 60-70 metres of strata built up [something that by any evolutionist's scenario would take at least millions of years] on the discarded knee-joint."[26]

In summary, it should be evident that the theory of evolution was advanced in spite of overwhelming and staggering evidences against it, with the proviso that evidence would be soon forthcoming. Some scientists set out to find the missing evidence—already convinced of the theory's veracity. Predictions were made that would have and should have been able to falsify the theory. The predicted evidence is missing in droves. The theory was then changed to explain the missing evidence, and the missing evidence (which was originally submitted by Darwin himself as a way to test or disprove his theory) is now used as a validation of the theory that apparently, needs no evidence. That alone should be enough to convince any unbiased observer that evolution theory is not a "science"—it is equally at home with evidence and the complete lack thereof. A theory that is able to explain everything and predict nothing should never be categorized as "science," much less as "fact"! As Malcolm Muggeridge once stated, "I myself am convinced that the theory of evolution … will be one of the great jokes in the history books in the future. Posterity will marvel that so very flimsy and dubious an hypothesis could be accepted with the incredible credulity that it has… I'm very happy to say I live near a place called Piltdown. I like to drive there because it gives me a special glow. You probably know that a skull was discovered there, and no less than five hundred doctoral theses were written on the subject, and then it was discovered that the skull was a practical joke by a worthy dentist in Hastings who'd hurriedly put a few bones together, not even of the same animal, and buried them ... So I'm not a great man for bones."[27]

Endnotes

1. For example, if a dinosaur bone could survive undecayed and unfossilized for sixty-five million years, any measurable C-14 would be gone 64.95 million years ago.
2. *Webster's New World Dictionary of the American Language, Second College Edition*. Cleveland & New York: William Collins + World Publishing Co., Ind., 1976, 275.
3. Darwin, Charles. *The Origin of Species. Great Books of the Western World, Vol. 49*. Encyclopedia Britannica, Inc., 1952.
4. Hitching, Francis. *The Neck of the Giraffe: Where Darwin Went Wrong*. New Haven and New York: Ticknor & Fields, 1982, 57-61.
5. Darwin. 152.
6. Raup, David M. "Conflicts between Darwin and paleontology," *Field Museum of Natural History Bulletin*. Vol. 50(1), (January 1979): 22.
7. Denton, Michael. *Evolution: A Theory in Crisis*. Bethesda, Maryland: Adler & Adler, 1986, 160-62.
8. Darwin, 309.
9. Stanley, *Macroevolution*, 1979.
10. Patterson, Dr. Colin. Letter written to Luther D. Sunderland. (April 10, 1979) cited in Sunderland, Luther *D. Darwin's Enigma*. San Diego, California: Master Books, 1984, 89.
11. Denton, 193-194.
12. Ibid., 189.
13. Virchow, Rudolph. Cited in Hitching, 208.
14. Chesterton, G.K. Cited in Hitching, 207.
15. Hitching, 208.
16. It is interesting to note that the "Piltdown Man" skull-fragments were somehow estimated to be five hundred thousand years old. When the fraud was eventually exposed, and the skull fragments were dated using carbon-14 methods in 1959, the carbon-14 dates yielded only hundreds of years (520-720). Such a discrepancy should be quite disconcerting to anyone uncritically accepting the vast age estimates for various fossils, especially considering that a true fossil cannot be tested by carbon-14 methods.
17. Osborn, H. F. Cited in Hitching, 211.

18 Larson, Edward J. *Summer for the Gods: The Scopes Trial and America's Continuing Debate Over Science and Religion.* Cambridge, Massachusetts: Harvard University Press, 1998, 32.
19 Ibid., 3-266.
20 Hitching, 212.
21 Lubenow, Marvin L. *Bones of Contention: A Creationist Assessment of Human Fossils.* Grand Rapids, Michigan: Baker House Books, 1992, 60.
22 Cuozzo, Jack. *Buried Alive: The Startling Truth about Neanderthal Man.* Green Forest, Arkansas: Master Books, 1998.
23 Ibid., cf. especially chapters 16-18 & 31.
24 Cremo, Michael, and Richard L. Thomson. *The Hidden History of the Human Race.* Badger, California: Govardhan Hill Publishing, 1994, 260.
25 Ibid., 261.
26 *Creation Ex Nihilo*, Vol. 11, No. 2, 18.
27 Muggeridge, Malcolm. *The End of Christendom.* Grand Rapids, Michigan: Eerdman's, 59.

Chapter 9

THE 2000 SEARCH FOR NOAH'S ARK

In 2000, as well as in 1999, I was privileged to do some glacier training at Palisade Glacier in the Sierras of California with some members of the Ark Research Project—ARP (including, but not limited to, Matthew Kneisler, Tim Mills, and Michael Holt). The leaders and trainers who gave of their time were Glenn and Edith Pinson, a couple from Southern California. I was also privileged later on to spend several days at Mt. San Jacinto in the snow and ice with Glenn and Edith teaching me the finer points of rock- and ice-climbing. (One of my most memorable backpacking experiences remains the Passover Seder that Glenn, Edith, and I shared together at San Jacinto in the snow.) I doubt that all of my combined training ended up qualifying me as anything but a novice "mountaineer," but at least I was no longer completely ignorant. I was not sure of Mustafa's (my Kurdish connection who offered the previous year to take me to the ark "for no money") qualifications for glacier climbing, so I felt I had to be as ready for any contingency as I could. During the course of the year, I sent him a letter in both English and Kurdish, but I got no reply. I also made several attempts to reach him via his cell phone number with no luck.

As August approached, it was clear that no one—group or individual—would be given any kind of official permission to search Ararat, so I was anxious to be able to reach Mustafa by phone. Finally, in July, the phone connected and I was able to speak with him at length. The call was not entirely encouraging, though he did remember his promise and seemed willing to follow through if conditions were right. However, less than two weeks before I was set to leave, I called once again and this time he said it was impossible for him to climb with me. He needed to leave on some kind of business out of the country, but he also seemed to feel quite strongly that the area we needed to access was still covered with too much snow to be able to photograph any objects present. My tickets and schedule were already set, and since I was returning to a teaching job that had already given me the first two weeks off, I was unable to go later in September. I decided to go and take my chances.

When I arrived in Dogubayazit, Paul Thomson and Richard Bright were already climbing Mt. Ararat from the western side. They had arranged some kind of local military permission and were able to access parts of the Ahora Gorge from the west, as well as some of the glacier above the gorge. I met John McIntosh in Dogubayazit the first night while he was awaiting news from Paul and Dick. It also happened that I encountered Mustafa and he now said that one reason he could not climb was that he had a knee injury that precluded climbing. He did, however, agree to meet me at the hotel later that evening. When we met, I asked if I could pray for his knee and allow Jesus to heal it. He was somewhat hesitant, but he agreed. The next day, he claimed that his knee was no better. Whether it was even injured in the first place, or whether God chose not to heal it, I do not know, but in any case, that door seemed closed for 2000.

Consequently, John and I waited for two more days until Paul and Dick had returned. They were able to get many photos as well as a video of the Ahora Gorge, and some of the upper glaciers. They had not seen any obvious objects of interest, but we nevertheless spent several hours poring over both the pictures and the videos trying to identify any potential locations. One sight in particular intrigued me as a possible location for the Ed Davis sighting. As previously mentioned, Ed claimed that in 1943 he was taken by Lors from Iran to Mt. Ararat, where they showed him two parts of the structure of the ark. At the time, he claimed, he could see two large pieces of the ark, broken open and lying on their sides. He claimed that he viewed the structures from above, looking down into a horseshoe-shaped valley through which streams ran before going over a waterfall. He voluntarily submitted to extensive lie detection analysis, which he passed with flying colors, and I have always felt his eyewitness account was among the most credible of the many claimed ark sightings.

However, there were aspects of it that seemed irreconcilable with other accounts placing the ark high on the mountain, on a ledge, and at least partially covered by ice. Yet, looking at the video Paul took, the top of what is called Avalanche Canyon looked like a very plausible location for the identifying features Ed had referred to. There even seemed to be two large rectangular structures buried under rubble and debris that could have been the sections that Ed saw. If so, they are fully covered by dirt and debris from avalanches and would take considerable excavations to access. Another portion of Ed's testimony, which I had previously overlooked or ignored, was that he was told that the ark was broken into "two or more" major pieces. If so, the upper piece could be the part claimed by other eyewitnesses to sit "on a ledge"—or it could be that most of the ark was once at this upper location, but is no longer. In any case, it seems plausible that the Davis pieces do not coincide with other claimed sightings because they are different sections and locations.

Paul, Richard, and I still had some time to spare so we spent several days pursuing options for reclimbing the mountain. Though there was still no official governmental permission to be had, some foreigners had gotten local military permission to make "sport" climbs from the south. Though this would not suit our purposes, it did appear that military oversight was more lax than it had been in recent years. Consequently, some friends in a Dogubayazit tourist office felt they could get us local permission for another climb from the west. Our passports were copied and faxed and we were told to wait hopefully for "one day, maybe two." Unfortunately, several "one days" came and went until Paul was out of time and had to return to New Zealand. Meanwhile, I read all the books I had brought with me—and waited. We did actually meet an official from the town of Igdir (on the west) who assured us that he could give his permission for a climb, but it would have to await the end of the weekend (Monday, the twenty-eighth of August, which was too late for me, as I had to leave on the thirtieth).

In any case, I had decided it was not going to happen and was making plans to return to the U.S. early when I received a call from Richard on Friday evening saying, "Get your pack and meet me within the half hour; we're leaving tonight." Local friends had arranged a four-wheel-drive vehicle and we were to be driven to Lake Kop on the western side of the mountain. Doing so would potentially save at least a day's climb. Eventually, it was determined that our "permission" was tenuous enough that it would be better to leave early in the morning when our vehicle was not so noticeable to any military observers. Richard, who is a good ten years my senior, had decided to hire a local Kurd to carry the bulk of his pack and climbing equipment. In retrospect, I would have been happy with the same arrangement!

We left Dogubayazit somewhere around 4:30 a.m. while it was still dark and were driven around to the west to a dirt road that was once utilized by the military in their fight against the PKK. Our driver eventually decided he was unable to take us all the way to Lake Kop, and he dropped us off about a mile or two from our hoped-for destination. That extra "mile or so" proved to be significant. Our goal had been to make our way around to the north and climb to a high camp on the first day. From there, we could hike over the ice to the area known as the "Steven's site" on the second day. If time permitted, we would then access the upper portion of the eastern ridge of the Ahora Gorge and try to take as many photos from there as possible before heading back to high camp. We would then make our way back to Lake Kop on the third or fourth day and call for a return ride on Monday or Tuesday at the latest.

The young man, Toucan, whom Richard had paid to act as porter was a very strong climber and immediately started up the western slopes. We were headed directly toward the western glacier and the feature known as the "Eye of the Bird," or the "Ice Cave." I had never seen such significant melt-back of the snow and ice. The entire area around the ice cave was devoid of snow and ice—normally the feature is barely visible through its ice covering. It was certainly readily apparent that there is no ark in this area.

After a couple of hours of climbing, we began to be concerned that we were heading too far to the west and would not be able to access the northern climbing route that we had planned. And now our guide and porter's inability to speak English became a greater concern. We thought we had clearly arranged our route through an interpreter before leaving town, yet now, off the path we had laid out, we seemed unable to get Toucan back on track. Initially we assumed he knew a way around to the north that we did not. But, after we'd climbed for a couple of hours, it became clear that he had no intention of taking us where we wanted to go. Whenever Richard addressed the question of our planned route, he would say, "Impossible." When pressed, he would attempt to say there was a military encampment on the north and we could not go that way. Somewhere around 9:00 a.m. as we caught up to him, he was in the process of setting up camp for the day! Our goal had been to climb the entire day and access high camp on the north side. We used the cell phone to call our contact in town and try to iron out the differences, but it was to no avail. Toucan was determined to go no farther. In retrospect, I would have to say he was both lazy and dishonest: he simply had no intention of spending the entire day climbing.

In any case, Richard angrily took his own pack and supplies back from Toucan, told him to go home, and we then decided that to turn back would waste most of our day—and potentially short-circuit our attempt completely. We therefore decided to push on and attempt to climb the western glacier and, once on top, make our way over the mountain to our destination. The climb proved both hazardous and extremely difficult as the mountain was steep and the terrain consisted primarily of much loose rock and debris. Many times a step would produce a small rock slide and it seemed that for every three steps forward, we often took two steps backward just trying to maintain footing. Eventually, however, we came to a vertical drop-off that was impassable. Though we could have set up an anchor and rappelled down the cliff, it would have left us without rope on the rest of the climb and on the ice, and our return route would have been difficult if not impossible. Furthermore, as we neared the snowfield at the base of the western glacier, we realized how steep and difficult a climb it would be—especially carrying all our gear. Unable to get down or around the cliff in front of us, we decided the only recourse was to reverse direction and try to find an alternate route around to the north.

It was not to be. Everywhere we turned, we were eventually forced back in the direction we had come. We would begin to make progress only to come to another impassable cliff. The route we ended up following as we attempted to descend seemed even more severe and difficult than our initial climb. I cannot even estimate how many times one or the other of us lost our footing and fell. I do remember that I began praying fervently that neither of us would sprain an ankle or break a leg. I knew that if Richard broke an ankle, I would undoubtedly have to leave him and go for help—I could not carry him by myself.

Late in the afternoon we finally stopped to cook some freeze-dried stew and have our first meal of the day. Later, by the time we came to set up camp, it was already nearing dusk and we had not even made it back to our starting point for the day. That night it rained considerably and the wind blew incessantly—and I discovered, much to my chagrin, that my bivy sack was not waterproof at all! In the morning as we assessed our position, we realized that, carrying our full packs with all our climbing gear and ropes, we did not have enough time to get back on the right path and still make our goal. Reluctantly, we realized we would have to abort our mission and return to town. John McIntosh had given me his cell phone before he left for America and I had taken it with me on our climb to use in an emergency. I had assumed that we would be using Toucan's phone, and mine was only along as backup. Neither Dick nor I had foreseen the possibility that Toucan would leave and take his phone with him. The batteries on my borrowed phone were low and several earlier attempts to call town had been to no avail. We now began to face the prospects of having to climb all the way down the mountain with little or no water. We were below any streams and neither of us liked the prospect of trying to find and approach a shepherd's camp to ask for water. Without adequate water, we could not fix any meals because our only food—with the exception of a few candy bars—was freeze-dried. I felt we could manage for several days without food, but it would be dangerous if we ran out of water.

We took a break in the early afternoon in order to have a candy bar and rest, and in the process of rearranging his pack, Richard came across another liter of water. That brightened our prospects considerably, but we decided to make the phone an object of prayer and try it one more time. It then occurred to me that there was a tourism office in town where a friend of mine spoke English. Miraculously, in my pocket, I found the number for the office, prayed again, dialed the number, and the call went through. I was able to ask my friend to call our "contact" and request a pickup. The phone worked for that one call and no more. However, we were not sure if the contact could be made, or if made, if anyone would be able to get to us since we were a day early. Consequently, we knew we had to continue our climb as if returning all the way to town. That meant conserving water, but it also meant that we could not take a shorter route by leaving the dirt path that served as a road—if we left the road by too great a distance, we would miss the car potentially coming for us. Yet, if there was no car, staying close to the road would mean adding several miles to our climb. As it was, we decided to take shortcuts where it was appropriate, as long as we thought we would be back to the road before a vehicle could approach unseen. In the end, we still almost missed our ride! We had just risked one long cutoff when I came to the road and took off my pack for a few moments to await Richard (and look for one last candy bar!) when the car arrived! Richard was not yet at the road. Had the car come just a couple of minutes earlier, we would have missed it entirely, and it might have been many hours before they realized they had missed us and came back looking for us. I don't think I have ever been so grateful for either a ride or cold water (with the exception of my 1999 return climb!).

I do recall that as Richard and I had been setting up camp the evening before, and assessing our climb, we had both sworn never to return (of course, I had heard that from Richard before, but never had he seemed so serious). In fact, I tried to think of some deep, dark secret about my life that I could tell to Richard so that he could threaten to publish it if I ever considered returning for another attempt at Ararat. In the end, however, neither of us could think of anything bad enough to tell the other that would be any worse than our memory of the climb down. It sounds overly dramatic in retrospect, but at the time, it seemed perfectly fitting to recall Paul's words to the Corinthians: "We do not want you to be misinformed, brothers, about the hardships we suffered in [Turkey]. We were under great pressure, far beyond our own ability to endure, so that we despaired even of life. Indeed, in our hearts we felt the sentence of death. But this happened that we might not rely on ourselves but on God, who raises the dead! He has delivered us from such a deadly peril, and he will deliver us. On him, we have set our hope that will continue to deliver us, as you help us by your prayers [II Cor. 1:1–8]."

I still fully believe that Noah's Ark is preserved on Mt. Ararat in eastern Turkey and that it has been preserved for an unbelieving era such as ours. There is a man named George Stevens III who does interpretations of remote-sensing, infrared satellite photo data for the military. He has identified at least two large, rectangular structures in the ice above the Ahora Gorge on Ararat. He believes with 100 percent certainty that they are man-made of organic material—possibly wood—and were once joined (interestingly, he does not believe in Noah's Ark, but he cannot imagine how any man-made structure was ever brought to this location). They are sitting in a fault and on a ledge and both buried by much ice. When he first identified the upper piece, he claimed there was about seventy feet of ice covering it.

In the past, I have had a difficult time reconciling his "discovery" with the Ed Davis account, because everything about them seems different. If the ark was indeed covered by seventy feet of ice, it would have been only under very rare weather conditions that it was accessible again. Indeed, this summer, after several seasons of drought and melt-back conditions, Paul and Richard were able to access the area of the "Stevens site" and found no structures visible. In any case, it has seemed to me that Ed Davis and George Stevens cannot both be correct—until this year. I believe that I can now reconcile the two accounts with something else that Ed Davis claimed, and that is that Lors told him that the ark was in three or four pieces. I think there was a part of me that still hoped the ark was mostly intact—or that it was in only the two pieces that Ed Davis saw. However, it now seems possible to me that the ark was once intact at the position of the upper Stevens site. If it sat on "a ledge" at this location, it would usually have been covered by snow and ice. Yet, as the ice melted in summer, one end of it would have periodically become visible, thus accounting for many of the sightings. However, if the ledge broke away, the portion of the ark still encased in ice would remain there, and the other section could very well have moved down the mountain and broken apart further in the process. That would mean that the portion that remained above would generally be fully covered in ice (since it is smaller and does not protrude as far as it did) and undetectable. It is interesting that, after my return from Turkey this year, I spoke with another friend and ark researcher, Robin Simmons, who told me that George Stevens now thinks the ice-melt conditions will make the upper portion of this object visible once again by late summer of 2001. That just may mean one more attempt after all.

Chapter 10

THE MYTH OF THE ICE AGE

If there was a global flood approximately 4,500 years ago, then it should be abundantly clear that any reconstruction of the history of our planet or its human inhabitants that attempts to bypass or ignore the flood is not just untenable, it is ludicrous. And yet, not a single history book in any public school in America—at any level—even mentions in passing the Noahic Flood. On the other hand, virtually all are replete with stories of one or more "ice ages"—something totally unheard of before the middle of the nineteenth century. Many Christian historians and scientists have even begun seeking ways to insert the "Ice Age" into the biblical record of history (where no secular scientist would ever put it, and where no theologian ever "found it" prior to the advent of evolution theory and its requirement of vast eras of time). I believe that the story of the ice age is fiction—or "myth." But it is a dangerous myth that at its roots was designed to eliminate belief in the biblical account of history in general, and the Noahic Flood in particular.

One of the things that seems to be missing in current literature pertaining to the "Ice Age" (Creationist literature included) is an examination of, or understanding of, the roots of the entire theory. The "discovery" of the Ice Age, like the invention of the theory of evolution, was not based upon scientific necessity, but the desire to extend time. In fact, the history of the formulation of the Ice Age theory is replete with antibiblical philosophy and uniformitarian ideas—not science. And though the story of a global flood is found in some form in virtually every culture in the world, there is not a single, tangible reference to continental sheets of ice taking over large portions of the world in any source of history. Creationist author

The Myth of the Ice Age

Michael Oard theorizes that the "Ice Age" was actually a post-flood, semicataclysmic, rapidly moving "ice age," complete with bursting of dams and destruction of vast areas of wildlife (including the woolly mammoths of Siberia and Alaska).[1] I would counter that if such a thing had occurred, history could not avoid mention of it. After all, if millions of mammoths had managed to proliferate after the flood, then the world would similarly have been refilled with people, as well as dinosaurs of all sorts. Tropical forests (in Siberia, Alaska, and elsewhere), according to Oard's view, had to have regrown quickly after the flood, and hundreds of thousands (or millions) of mammoths—and numerous other animals—had repopulated this region, only to be redestroyed in another post-flood cataclysm. Yet there is no hint of such a thing in any historical reference anywhere. Especially the Bible. Such an omission is hardly tenable. The historical reality is that the idea of the "Ice Age" has far more to do with the idea of uniformitarianism than it does with science or evidence.

Historically, one of the first naturalists to attempt to advance the idea of the earth being covered by continental ice sheets was Louis Agassiz. (Interestingly, he personally saw an "ice age" as a rapid and cataclysmic inundation of the earth. But his cataclysmic view was not to hold sway for long.) In 1836, he built a hut on the edge of a glacier and lived there in order to make his observations. Unfortunately, even though his observations were initially limited to one small area of Switzerland, he made conclusions about the rest of the world. He was able to convince Buckland (a believer in the Noahic Flood) and in 1840, they both presented papers before the Geological Society of London.[2] According to Buckland, he had written to Agassiz on the success of his presenting their new "ice age" theory to a young man named Charles Lyell: "Lyell has adopted your theory in toto!!! On my showing him a beautiful cluster of moraines, within two miles of his father's house, he instantly accepted it, as solving a host of difficulties that have all his life embarrassed him."[3]

It apparently took little effort to convince Lyell of the concept of glacial ice sheets covering continents, but what was it that had "embarrassed" Lyell? Velikovsky points out that Lyell realized that floating icebergs were an inadequate explanation of the "phenomena of drift and erratic boulders in all places. The only alternative had [previously] been the waves of translation, or tidal waves traveling on land, but this was outright catastrophic."[4] In fact, glacial ice sheets were far more agreeable to Lyell primarily because they not only eliminated explanations that involved catastrophic floods, but they potentially added huge amounts of time to earth's "history"—by his estimate, at least one million years.[5]

Scientifically, Lyell did not even need to go around the world and examine the nature of the evidence, because the theory of the Ice Age was to him a godsend (no pun intended). In retrospect, it would seem that Buckland apparently did not recognize that the aspect that appealed to Lyell was *time*: the idea of cataclysmic, continent-wide destruction having been caused by a huge flood could be supplanted by the alternative of a continental sheet of ice moving slowly over the millennia. As Lyell would later write in *Principles of Geology*, "All theories are rejected which involve the assumption of sudden and violent catastrophes and revolutions of the whole earth, and its inhabitants ..."[6] Of course, he might as well have said, "The biblical account of history, and especially the account of the Noahic Flood, is rejected as nonsense, and all evidence heretofore seen as supporting it must now be reinterpreted."

As Lyell would also write, "Conebeare [geologist and Bishop of Bristol] admits three deluges before the Noachian! And Buckland adds God knows how many catastrophes besides, so we have driven them out of the Mosaic record fairly."[7] The battle was not so much for the existence of an "ice age" as it was against the historicity of the Bible. Lyell recognized, as many modern theologians apparently do not, that the individual battles were not as crucial as the War: it was the foundation of truth

The Myth of the Ice Age

that counted. If he could get men to recognize the authority of man's interpretation of geology over and above that of Scripture, he had won the war. It would not matter if certain individuals still believed in a worldwide flood, if they would concede the authority of geology to establish true history (even if they tried to find a way to insert that "history" into the Bible), the authority of Scripture had been usurped. And eventually the Bible and the account of Noah would be relegated to the scrap pile of useless fairy tales. In most of society, and alarmingly larger segments of Christianity, that is right where we find ourselves today.

Yet, even in the mid-1800s when the idea of "ice ages" was beginning to catch on, there were dissenters who looked at the evidence through a very different interpretive lens. R. I. Murchison, for instance, went to Russia for a geological survey and found that large erratic boulders strewn over the plains of Russia (attributed by Agassiz to distribution by an ice sheet) got smaller in size the farther south he went. He argued that this instead pointed to the cataclysmic action of water—a massive flood.[8] Murchison opposed Agassiz, Buckland, and soon thereafter, Lyell. He argued that the sand, stones, clay, and gravel, which were dispersed over much of Russia and Europe, had "been transported by aqueous action, consequent of powerful waves of translation and currents occasioned by relative and often paroxysmal changes of the level of sea and land."[9] He insisted that "aqueous detrital conditions will best account for the great diffusion of drift over the surface of the globe, and at the same time explain the very general striation and abrasion of the rocks, at low as well as high levels, in numerous parallels of latitude."[10] The regions of Russia, Finland, and Sweden that were powerfully scoured held no mountains from which any glaciers could have issued. In fact, that is the case worldwide with much of the so-called ice age evidence. Indeed, in 1865 when Agassiz traveled to Brazil, he saw there the same "evidences" that had led to his conclusion of ice sheets elsewhere. There were scratched rocks, erratic boulders, and the like, but Brazil is at the equator, as is

British Guyana and equatorial Africa and Madagascar. And in many of these locales, including India, it was necessary, because of the direction of striations or deposits, to postulate that the ice sheets spread from the equator, even when there were no mountains available for the gravitational push—or the push came toward the mountains.[11] All in all, an entirely unlikely (read: impossible!) phenomenon.

Existing awkwardly in the midst of this new explanation of earth's history—entirely devoid of any real history—are the woolly mammoths of Siberia—at least some of them seemingly flash-frozen into the tundra, complete with the remains of semitropical plants in their mouths and digestive systems. These were so troubling to Lyell that he, in essence, simply chose to ignore them. Darwin, in following Lyell by denying the occurrence of any continental cataclysms in any era, nevertheless admitted in writing that "the extinction of the mammoths in Siberia was for him an insoluble problem."[12] Darwin based his ideas of evolution and slow, gradual change over millennia in large part on his acceptance of Lyell's principles of geology and the concept of uniformitarianism. At one point in *Origin of Species*, he declares that if a person can read Lyell's work on geology and not accept the concept of vast time on earth, he need not bother to read his own work any further. His "work" depended absolutely on the precepts established by Lyell.[13] And Lyell's "work" was to replace the biblical flood with an "ice age." The evidence of the woolly mammoths should be seen as overwhelming support for the global flood of Noah.

The idea currently in vogue in some creationist circles that the mammoths of Siberia and Alaska were somehow destroyed in a post-flood "ice age" needs to be challenged both biblically and geologically. The concept that vast, continent-wide destruction occurred apart from—or centuries after—the flood cannot be supported by the biblical account of history. No biblical scholars working exclusively with Scripture ever discerned the existence of an "ice age." Such an idea does not exist in Scripture

The Myth of the Ice Age

even between the lines. The destruction that overtook the mammoths was vast, cataclysmic, and extremely rapid—namely the Flood of Noah. The mammoths were buried in situ with semitropical plants in their mouths and digestive systems and their flesh was rapidly frozen, directly implying that Siberia was a warm, plush, environment up until the moment of the cataclysm that overtook them.

As F. C. Hibben wrote in 1943, "Although the formation of the deposits of muck [in Siberia] is not clear, there is ample evidence that at least portions of this material were deposited under cataclysmic conditions [note: it is not inherently possible that "portions" of a vast layer of muck were deposited cataclysmically and the rest was not, nor is it possible that it was gradually transported by glacial ice]. Mammal remains [found in the "muck"] are for the most part dismembered and disarticulated, even though some fragments yet retain, in their frozen state, portions of ligaments, skin, hair, and flesh. Twisted and torn trees are piled in splintered masses."[14] Furthermore, as Immanuel Velikovsky points out, no matter what else happened, it is essential to comprehend that only a hurricane or flood, or combination of both, could have produced the type of destruction that is evident on a continental scale.[15] This is clearly not the result of a moving sheet of ice, no matter how rapid one tries to make it.

In an issue of *Creation* magazine, the significance of the preservation of stomach contents of some Siberian mammoths was downplayed by comparing it to an unfrozen, mummified "mammoth" (later corrected to mastodon) found in U.S. soil that also had preserved stomach contents.[16] However, such a comparison is moot: it is not the state of the stomach contents that is of import, it is the nature of the stomach contents and the state of the flesh of the mammoth that cry out for explanation. The nature of the stomach contents of the Siberian mammoths indicates an environment that was warm and plush—completely unlike the environment of Siberia today. Not even a "rapidly moving" ice sheet or ice age could in any way account for a woolly mammoth, in the process of chewing its bean pods—which only grow in warm environments—being overwhelmed by glacial ice! Second, it is the actual flesh of the mammoth that is astonishing in its implications—it is nothing like the unfrozen, mummified, mastodon of North America. There are persistent, though unsubstantiated claims that the flesh of some unearthed mammoths was actually consumed by humans—no one could even pretend that the mummified mastodon of North America was edible. The flesh of some of the mammoths of Siberia was so rapidly frozen that the cellular tissue had not been broken down. This would be impossible even for a cut of fresh beef placed into

Berezovka Mammoth found in 1901, preserved in Zoological Museum, St. Petersburg, Russia. Portions of his head and trunk had been eaten by predators before scientists arrived to excavate his remains. Undigested plant remains were found in his mouth and stomach, demonstrating the extreme rapidity of his burial, as well as the fact that the climate at the time of his death was vastly different than the climate of Siberia today. [Fig. 132, *In the Beginning: Compelling Evidence for Creation and the Flood*, 8th ed., 2008, by Walt Brown, p. 228, from Zoological Museum of St. Petersburg.]

a freezer of today.[17] The stomach contents may or may not be preserved under different circumstances, but the flesh of some of the mammoths frozen into the tundra of Siberia indicate that they were buried cataclysmically and frozen virtually instantaneously, and they have remained that way until rediscovered in modern times. There is no viable alternative.

Articles and comments in such venues as *Creation* magazine have also tried to downplay the numbers of mammals found in this condition, but the finding of even one is inexplicable by any "ice age" hypothesis, no matter how rapid. The actual finds, dispersed over an entire continent, undermine completely any theory that does not involve a continent-wide, rapid, and immediate cataclysm. And the numbers are, indeed, large. Walt Brown cites fifty-eight known reports of mammoth and rhinoceros discoveries as well as the discovery of the fleshy remains from many other animals.[18] In the 1970s, Russian scientists removed roughly nine thousand mammoth bones from one burial site alone.[19] According to Nikolai Vereshchagin, chairman of the Russian Academy of Science's Committee for the Study of Mammoths, "more than half a million tons of mammoth tusks were buried along a 600-mile stretch of the Arctic coast."[20] According to most estimates, such an amount would require the existence of over five million mammoths in this region alone. Even if this were postulated to be the accumulation of thousands of years' worth of mammoths, this could not possibly have taken place postflood. Sir Henry Howorth summarized the mammoth dilemma succinctly: "The instances of the soft parts of the great pachyderms being preserved are not mere local and sporadic ones, but they form a long chain of examples along the whole length of Siberia ... so that we have to do here with a condition of things which prevails, and with meteorological conditions that extend over a continent.

"When we find such a series ranging so widely preserved in the same perfect way, and all evidencing a sudden change of climate from a comparatively temperate one to one of great rigour, we cannot help concluding that they all bear witness to a common event. We cannot postulate a separate climate cataclysm for each individual case and each individual locality, but we are forced to the conclusion that the now permanently frozen zone in Asia became frozen at the same time from the same cause."[21]

Most current secular science accounts insist that the mammoth was an animal suited for cold-weather environments (primarily because of its hair, and the fact that its remains have been unearthed in the frozen tundra of Siberia).[22] Yet such an interpretation overlooks the remains of the many other animals found in the same regions, which include the antelope, camel, tiger, beaver, woolly rhinoceros, badger, leopard, elk, bison, and numerous others—most of which are not found in cold-weather regions today.[23] In addition, the actual physical characteristics of the mammoth indicate anything but a cold weather animal. In fact, in perhaps the most detailed study yet conducted of the mammoth skin and hair, the author concluded, "It appears to me impossible to find, in the anatomical examination of the skin and [hair], any argument in favor of adaptation to the cold."[24] Among the evidences cited against the mammoth existing in cold climates are the facts that its skin lacked oil glands (something present in all of today's Arctic mammals); elephants, and therefore presumably mammoths, must be kept quite warm in order to survive (especially the young); layers of fat under the mammoth's skin indicate an abundance of food; mammoths, like current elephants, would have required large amounts of drinking water—unavailable in the freezing Arctic; the long hair of the mammoth would actually have been a detriment to survival in cold climes because snow and ice would have clung to the hair, accumulated, and eventually been harmful; and furthermore, the dietary requirement of salt would have been virtually impossible to come by in the frozen tundra.[25]

The Myth of the Ice Age

The specious argument that few mammoth remains have actually been unearthed in a perfectly preserved state is of little or no merit because much of this area is very sparsely populated and most finds go unreported for various reasons, especially discoveries that would delay or interfere with mining prospects of the region. In addition, ivory tusks have been traded extensively from this area for almost four hundred years, during which an estimated one hundred thousand tusks have been exported. Tusks exposed to the elements, unless kept frozen, deteriorate and crumble.[26] The fact of the matter is that moving glaciers—slow or rapid—could never account for the condition of the mammoths that have been documented, let alone their numbers. In 1846, a surveying team found a newly exposed mammoth (uncovered by overflowing rivers), so perfectly preserved that they first thought it to be living. When they were eventually able to pull it to shore, they cut it open and they found its stomach full of chewed fir cones, and shoots of fir and

Illustration of the "swimming" Benkendorf Mammoth. [Fig. 137, *In the Beginning: Compelling Evidence for Creation and the Flood*, 8th ed., 2008, by Walt Brown, p. 234, Steve Daniels.]

pine.[27] It had obviously been living in a plush and temperate climate at the moment it was rapidly inundated and permanently frozen within an extremely short period of time.

In addition, the condition of many of the finds, not exclusive to mammoths, bespeaks not just rapid inundation but catastrophic conditions as well. The famous Berezovka Mammoth (found in an eroded riverbank in 1900 and partially eaten by dogs) was upright, but extremely contorted with many broken bones. As Walt Brown points out, the long bone in his right foreleg was crushed into numerous pieces without even damaging the surrounding tissue. For this to happen, his leg must already have been encased in sediment while tremendous weight was applied vertically. He appeared to have suffocated while in the very process of eating grasses and bean pods, some of which remained unswallowed and even unchewed! Several other finds in the same area, including a young mammoth and rhinoceroses, also showed signs of death by suffocation and severe exertion before death.[28] In addition, twenty-four pounds of undigested food was removed from the stomach of the Berezovka Mammoth, and much of the plant life contained therein cannot grow in current Arctic conditions. It is abundantly clear that this portion of the world was a temperate and verdant ecosystem that was home to countless animals when it was catastrophically overwhelmed by huge amounts of airborne and waterborne sediments, and that much of it rapidly froze in the aftermath and has remained so ever since.

The cataclysm that overwhelmed the mammoths and many other animals in Siberia, Alaska, and much of the Arctic regions was the precipitating event of the Flood of Noah when "the fountains of the great deep" burst forth. Huge amounts of water were jettisoned into the atmosphere under tremendous pressure. The rains that fell included large amounts of silt and sediments, which were transported by high winds. This area rapidly froze and has remained frozen ever since. The buildup of ice in

The Myth of the Ice Age

the Arctic and the Antarctic was extremely rapid in the centuries following the flood, and these areas have never returned to their pre-flood conditions. Walt Brown succinctly summarizes it thus: "The rupture of the earth's crust passed between Alaska and Siberia in minutes. Jetting water from the 'fountains of the great deep' first fell as rain. During the next few hours, subterranean water that went above the atmosphere, where the effective temperature is several hundred degrees below Fahrenheit, fell as hail. Some animals were suddenly buried, suffocated, frozen, and compressed by tons of cold, muddy ice crystals from the gigantic 'hail storm.' The mud in this ice prevented it from floating as the flood waters submerged these regions after days and weeks. A thick blanket of ice preserved many animals during the flood phase. After the mountains were suddenly pushed up, the earth's balance shifted, the earth 'rolled,' so what is now Alaska and Siberia moved from a temperate latitude to their present position... As the flood waters drained off continents, the icy graves in warmer climates melted, and the flesh of those animals decayed. However, many animals, buried in what are now permafrost regions, were preserved."[29]

Though the preceding invariably contains speculation, it is based on not only the biblical account of history, but the available data, and is, I believe, the best-fit explanation.

There is not a shred of historical—or biblical—evidence for the existence of a plush, semitropical world in either Siberia or the Antarctic following the flood. In fact, there is no written record in any culture of the world for anything like an "ice age." The Ice Age is a myth that was perpetrated largely by those who were seeking to undermine belief in a global flood and faith in the veracity of the history recorded in the Bible.

Endnotes

1. Oard, Michael. *Frozen in Time: The Woolly Mammoth, the Ice Age and the Bible.* Green Forest, Arkansas: Master Books, 2004.
2. Velikovsky, Immanuel. *Earth in Upheaval.* New York: Doubleday & Co., Inc., 1955, 34-35.
3. Agassiz, Elizabeth Cary, ed. *Louis Agassiz, His Life and Correspondence.* 1893, 309, cited in Velikovsky, 36.
4. Velikovsky, 37.
5. According to Velikovsky, (footnote 6, 37) Lyell simply borrowed the guess of one million years from J. Croll who had advanced an "astronomical theory" of the ice age (that is no longer accepted). Again, the evidence of massive flooding was simply reinterpreted in order to extend time.
6. Lyell, Charles. *Principles of Geology, 12th ed.* 1875, 318, cited in Velikovsky, 27-28.
7. Lyell, cited in Velikovsky, 234-35.
8. Murchison, R. I. *The Geology of Russia in Europe and the Ural Mountains.* London, 1845, 553, cited in Velikovsky, 38.
9. Ibid.
10. Murchison, 554, cited in Velikovsky, 39.
11. Velikovsky, 40-41.
12. *Journal of the Philosophical Society of Great Britain, XII*, 1910, 56, cited in Velikovsky, 6.
13. Darwin, Charles, *Origin of Species. Great Books of the Western World, Vol.49.* Encyclopedia Britannica, Inc., 1952, 153. The actual quote is: "It is hardly possible for me to recall to the reader who is not a practical geologist, the facts leading the mind feebly to comprehend the lapse of time. He who can read Sir Charles Lyell's grand work on the *Principles of Geology*, which the future historian will recognize as having produced a revolution in natural science, and yet does not admit how vast have been the past periods of time, may at once close this volume."
14. Hibben, F. C. "Evidence of Early Man in Alaska," *American Antiquity, VIII.* (1943): 256, cited in Velikovsky, 2.
15. Velikovsky, 2.
16. *Creation*, 24(2): 22, and 24 (3): 5.
17. As Walt Brown points out, "When an animal dies and decay begins, decomposing amino acids in each cell produce water that ruins the meat's taste. Water expands as it freezes. If a cell freezes after enough water has accumulated, the expansion will tear the cell, showing that a certain amount of time elapsed between death and freezing. This characteristic was absent in the Berezovka mammoth, and the meat was edible—at least for dogs. Apparently, these mammoths froze before much decay occurred." *In the Beginning: Compelling Evidence for Creation and the Flood*, 7th

18 Brown, 161. Author's note: If this region could be systematically and thoroughly examined, I predict that the intact remains of various dinosaurs will also be discovered, and stomach contents of even the "carnivores" among them will show the remains of a vegetarian diet. See also chapter 12, "The Scriptural Advent of Animal Carnivorousness."
19 Stone, Richard. *Mammoth: The Resurrection of an Ice Age Giant*. Cambridge, Massachusetts: Perseus Publishing, 2001, photo pages between 114 & 115.
20 Stewart, John Massey. "Frozen Mammoths from Siberia Bring the Ice Ages to Vivid Life," *Smithsonian* (1977): 67, cited in Brown, 159.
21 Howorth, Henry H. *The Mammoth and the Flood*. London: Samson Low, Marston, Searle and Rivington, 1887, 89, cited in Brown, 166.
22 But see also account by Michael Oard, *The Mammoth and the Ice Age* (video/DVD), or his book, *Frozen in Time: The Woolly Mammoth, the Ice Age, and the Bible*.
23 Brown, 162.
24 Neuville, H. "On the extinction of the Mammoth," *Annual Report of the Smithsonian Institution*. 1919, 332, cited in Brown, 162.
25 Brown, 162.
26 Ibid., 164.
27 Ibid.
28 Ibid., 165.
29 Ibid., 169.

Note: The top of the page continues footnote 17 with: edition. Phoenix, Arizona: Center for Scientific Creation, 2001, 184.

Chapter 11

DATING A DINOSAUR: WHAT YOUR MOTHER NEVER TOLD YOU

Even from many Christians, I have heard the question, "How could dinosaurs have been on Noah's Ark if they were extinct millions of years before the flood?" And clearly, were the story of their demise roughly sixty-five million years ago, true, they could not have been on the ark. This chapter will attempt to place dinosaurs back into the historical context where they belong—both in the Bible, and on the ark. Sadly, apparently unbeknownst to many, the news of their demise "millions of years ago" is rooted and founded in evolutionary teachings and beliefs, but neither in history nor in science.

As established in earlier chapters (see chapter "The Spiritual Roots of Evolution Theory"), many of the early authors of evolution theory were diametrically opposed to the creation account of Genesis and absolutely required the extension of time in order to allow for evolution, as well as to denigrate the historicity of the Bible. Consequently, it has been essential to the concept of evolution to both establish the existence of vast eras of time, as well as to separate the fossil record into vastly different evolutionary epochs. If men, for example, had lived in "the age of the dinosaurs," then no evolution could have taken place. According to dinosaur expert David Lambert, it is an essential evolutionary precept that dinosaurs and man could not have existed anywhere near each other in time: "Contrary to comic books and motion pictures, no dinosaur survived to frighten early man 'one million years B.C.'"[1]

David Norman in his 1991 book *Dinosaur!* expresses a similar sentiment when he states: "Most people now

Dating a Dinosaur: What Your Mother Never Told You

know [sic] that dinosaurs became extinct about 66 million years ago, because it is a subject which has received considerable airing in the press, on television and the radio... The whole issue of time is further confused today through television and films. The very popular cartoon series *The Flintstones* gives the clear impression that Stone Age man lived alongside dinosaurs, and even had some as pets! [I don't know about you, but most of my perception of true history actually did come from watching the Flintstones!] It is difficult to erase the mistaken ideas that such scenes create. The simple fact is that dinosaurs vanished from the face of the earth almost sixty-six million years before modern humans appeared. Our history dates back a mere one hundred thousand years, so any thoughts of cavemen wrestling with Tyrannosaurus are completely nonsense."[2]

These statements show how essential it is for the evolutionist to separate dinosaurs from human beings—if they coexisted, according to the evolutionists themselves, the very concept of evolution is demolished. All the animals, including man and the dinosaurs, simply could not have coexisted or none of them could have been the ancestor of any of the others. Consequently, to demonstrate the lie of dinosaur existence (or extinction!) sixty-six million years ago is to destroy the foundation of evolution theory.

There are basically four areas or types of evidence to be examined: (1) Scientific evidence of the fossil record—if the fossil record shows evidence of man and dinosaurs existing at the same time, or if unfossilized dinosaur bones could be found, the implications of this fact alone should be sufficient to overturn any evolutionary precepts. (2) Modern accounts—though it is certainly not essential for dinosaurs to have survived until modern times, if they lived at the same time as all other creatures, and if they were therefore included on Noah's Ark, the likelihood certainly exists that some may still be around (unaltered by evolution). (3) Ancient historical accounts—if dinosaurs existed with man, we would expect to find mention of such creatures in the historical record, even if they were not named as "dinosaurs." (4) The biblical account—the Bible is either the source of truth and history in every area or it is untrustworthy in any area. And if dinosaurs existed with man, though it is not necessary that they be specifically singled out for mention in the Bible (as numerous other specific animals such as the platypus are not mentioned either), it is likely that the biblical account would support this claim.

(1) Scientific Evidence

In beginning to examine the scientific evidence of the fossils, it quickly becomes apparent that all evidence needs to be interpreted, and the interpretation depends upon beliefs and presuppositions. If one believes that dinosaurs form part of the evolutionary chain that came and went millions of years ago, evidence to the contrary will often be misconstrued. As Robert Jastrow (former director for the Institute for Space Studies, USA) once wrote, "The case of the disappearing dinosaurs is a fascinating demonstration that science is not based on facts alone. The interpretation of the facts is even more important."[3] In fact, overwhelming scientific evidence of the coexistence of dinosaurs and man is continually overlooked, ignored, or simply rejected out of hand. In addition, most dating methods are based, not upon testable science, but on the belief in evolution itself. Fossil dating is, in reality, a game in tautology. Dinosaur bones do not come with dates attached—none have been found with headstones marking the date of their demise (and if they were, one would suspect the headstone to have been erected by a human!). The scientific literature on fossils in general, and dinosaur fossils in particular, demonstrate two common features of fossils: (1) they often appear disarticulated and jumbled together with other animals as if torn asunder in cataclysmic conditions (see photograph below), and (2) they almost always appear to have been buried in an event associated with water. In fact, the truth is that fossils form in virtually no other way. For example, in the American Midwest it is estimated that set-

Dating a Dinosaur: What Your Mother Never Told You

A typical fossil bed where approximately 9000 different animals—including camels, wild boars, rhinoceroses—are entombed in one hill. This is only a portion of a once-larger deposit and clearly demonstrates the catastrophic nature of the event that buried and fossilized untold thousands of animals in this area. [Photograph from Nelson, Byron. *The Deluge Story in Stone*. Minneapolis, Minnesota: Bethany Fellowship, Inc., 1968, p. 99.]

tlers and buffalo hunters slaughtered millions of buffalo, and most of them were simply left where they were killed. To date, none of their remains has been discovered as fossils. Their carcasses rotted and were eaten by scavengers. Their bones, left to bleach in the sun, decayed and deteriorated and have essentially disappeared. In order to be fossilized, an animal must be captured, usually in waterborne sediment, buried quickly, and left undisturbed while the bones mineralize. (Note: see chapter "The Fossil Record" for more discussion of fossil formation and dating.)

It is pertinent to wonder why, if the large dinosaurs really died out millions of years ago, they are most often found in the uppermost layers of the strata, and why they most often seem to have died in profusion under cataclysmic conditions. Should they not be buried *under* the buildup of sixty-five million years' worth of strata? After all, that is how the fossil layers are said to have formed in the first place. A few quotations from various sources illustrate the way dinosaur bones are often found in the strata: (1) From a dig in New Mexico: "As the layer was exposed, it revealed a most remarkable dinosaurian graveyard in which there were literally scores of skeletons one on top of the other and interlaced with one another. It would appear that some *local catastrophe* [italics added] had overtaken these dinosaurs, so that they all died together and were buried together."[5] (2) From Alberta, Canada: "Innumerable bones and many fine skeletons of dinosaurs and other associated reptiles have been quarried from these badlands, ... a stretch that is a veritable dinosaurian graveyard."[6] (3) From a 1934 find in Wyoming: "The concentration of the fossils was remarkable; they were piled *like logs in a jam* [italics added]."[7] (Hmmm, when do logjams occur? It sounds remarkably like the aftereffects of a large flood.) (4) In the Morrison formation in Wyoming is an area known as the "Bone Cabin Quarry": "At this spot the fossil hunters found a hillside literally covered with large fragments of dinosaur bones that had weathered out of the sediments composing the ridge ... the party went to work, digging down into the surface of the hill, and as they dug, more and more bones came to light. In short, it was a veritable mine of dinosaur bones."⁸ (5) From 1909 to 1914 in Tanzania: "The site contained an enormous number of fossils—far more than could be carried off by one expedition. As in most of such sites, the greater parts of the remains were fragmentary ... there was much speculation as to how the remains of so many dinosaurs came to be concentrated in beds otherwise poor in fossil remains. Some German scientists suggested that the animals had been *overwhelmed by a natural catastrophe* [italics added]."⁹ (Or, more to the point, a Supernatural catastrophe if you consider that God initiated the flood!) These examples come from all over the world and they could go on and on. Remember, the one thing that all fossils have in common is that they are dead animals—killed by the very event that fossilized them! Where is that happening on earth today? The explanation for fossil graveyards that most scientists have tried desperately to avoid is that, instead of numerous, isolated, cataclysmic events spread over millions of years producing the fossilized dinosaurs, it was one cataclysmic event spread over the entire surface of the earth—namely, the flood of Noah only forty-five hundred years ago!

As previously stated, evidence or data must be interpreted. Even a precursory glance at the quoted examples of the various finds of dinosaur graveyards illustrates that the similarity alone could tie all of the finds together with one cause. Instead, the evidence is interpreted according to presuppositions and philosophy, and not according to any recognized principles of science.

A few years ago, a paleontologist named Mary Schweitzer examined the femur of a T.rex found in Montana. She "dated" the fossil at sixty-five million years old (Why? Because it's a T.rex, and that is when they died according to evolutionary theory!). However, she found strands of DNA as well as red blood cells inside the bone that was not fully fossilized. Knowing that such a thing would be absolutely irreconcilable with an age of sixty-five million years, she initially hypothesized that the DNA may have come from a fungus that invaded the bones recently. She stated: "Finding remnants of dinosaur blood cells would have astounding implications. Tiny bits of proteins and DNA possibly locked away inside the structures could contain the coded message of life just waiting for scientists to decipher them. Recently, the notion of finding preserved dinosaur DNA has produced a lot of headlines, not to mention blockbuster movies. Most scientists don't put much stock in the idea because it's unlikely that DNA could last for millions of years."[10] (Note: every laboratory experiment and every observation of nature and natural processes has demonstrated conclusively that blood cells and DNA could not survive decay processes for more than several thousand years. The idea that either could survive for millions of years is physically preposterous. So the implications of finding T.rex blood cells with DNA in them are indeed staggering! Such a discovery should conclusively demonstrate that dinosaurs survived until fairly recent times—or at least that one did!)

The blood cells found in the T-rex femur passed every test of modern science for proving them to be blood. As Carl Wieland summarized: "The tissue was coloured reddish brown, the colour of hemoglobin, as was liquid extracted from the dinosaur tissue. Hemoglobin contains heme units. Chemical signatures unique to heme were found in the specimens when certain wavelengths of laser light were applied. Because it contains iron, heme reacts to magnetic fields differently from other proteins—extracts from this specimen reacted in the same way as modern heme compounds. To ensure that the samples had not been contaminated with certain bacteria, which have heme (but never the protein hemoglobin), extracts of the dinosaur fossil were injected over several weeks into rats. If there was even a minute amount of hemoglobin present in the T.rex sample, the rats' immune system should build up detectable antibodies against this compound. This is exactly what happened in carefully controlled experiments."[11]

Dating a Dinosaur: What Your Mother Never Told You

How does the evolutionist respond to this? Usually with denial or even cover-up. As Mary Schweitzer said in her article: "I showed these microscopic slides to my boss, paleontologist Jack Horner, renowned for his work on dinosaur nesting sites. He took a long look and then asked, 'So you think these are red blood cells?' I said, 'No.' He said, 'Well, prove that they're not.' So far, we haven't been able to."[12] Notice that the underlying principle behind this type of research is not an open-minded search for the truth. If preconceived notions and philosophies were not already intact, the blood cells would be immediately hailed as great evidence for the actual "young" age of these bones. They could even be carbon-14-dated, and though C-14 dates are based on unreliable—and perhaps, demonstrably false—presuppositions, finding a date of thousands of years would necessarily shatter the myth of dinosaur fossils being untold millions of years old. Instead, the evidence is ignored, downplayed, or denied outright. In some cases it is even intentionally altered or destroyed.

Margaret Helder published an article in 1992 and stated of Mary Schweitzer, "She could not accept that fresh ... dinosaur bones had been found in Alaska. 'Such bones could never have lasted 70 million years,' [Mary] said. Unlikely or not, it is a fact that such bones have been found... How these bones could have been preserved for 70 million years is a perplexing question. One thing is certain: they were not preserved by cold. Everyone recognizes that the climate in these regions was much warmer during the time when the dinosaurs lived ... Why then did these bones not decay long ago? ... The obvious conclusion is that these bones were deposited in relatively recent times."[13]

In fact, Schweitzer and others divulged publicly in 2005 that when breaking open a fossil leg bone of a T.rex (in order to transport it better), they found it to be hollow and not fully fossilized. Inside, they found soft, flexible tissue as well as blood vessels that were still pliable (after sixty-eight million years!). Indeed, they reported that there still appeared to be blood inside the vessels and, in fact, blood could still be squeezed from some of the vessels! Both direct observation, as well as extensive experimentation, have shown these conclusions to be accurate.[14]

As Mary observed elsewhere, "It was exactly like looking at a slice of modern bone. But, of course, *I couldn't believe it*. I said to the lab technician, 'The bones, after all, are 65 million years old. How could blood cells survive that long?' [italics added]"[15] It should be clear that it is not the evidence that is "unbelievable," but the interpretation that is based on a faulty belief system.

In the early 1900s a man by the name of Charlie Moss reportedly found human footprints among dinosaur footprints in the limestone riverbanks of the Paluxey River in Texas. Many people claim to have seen them; some footprints were even cut out of the rock to be sold. This prompted charges that the prints had been carved as a hoax for profit. So, in the early 1980s Dr. Carl Baugh and others began searching for other corroborating evidence. There are many dinosaur prints left in the rocks in that area, so much so that a portion of the land has been set aside as "Dinosaur Park." In March of 1982, on land leased from a local resident who had both pictures and stories of his own finds, they began to find many apparently human footprints intermingled with those of dinosaurs. In some instances whoever or whatever left the footprints appeared to have been running as if to flee some common disaster. The limestone rock in which these footprints were discovered was in most instances buried by several feet of undisturbed sediment that had to be carefully dug away. Geologists affirm that the material in which the prints were found would have hardened like modern concrete within twenty-four hours of deposition, confirming that the prints found were undoubtedly made within hours or even minutes of each other. On Tuesday, March 16, 1982, a clearly defined human footprint was found only eighteen inches from a large dinosaur print—these were found underneath a layer of undisturbed limestone twelve inches thick.[16] In one ins-

Dating a Dinosaur: What Your Mother Never Told You

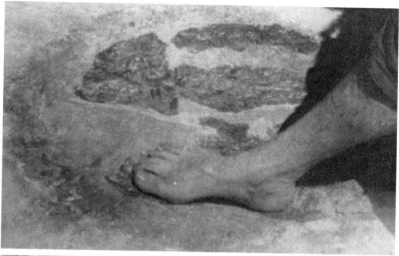

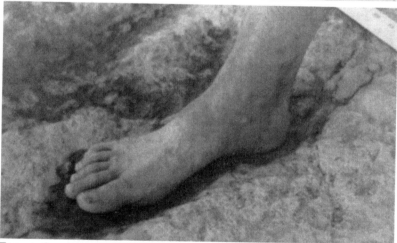

Top: a fourteen-inch human-appearing footprint adjacent to a dinosaur footprint. Bottom: Another fourteen-inch footprint found in summer of 1989 in what is referred to as the "Clark" site. [Photograph from Baugh, Carl E., & Wilson, Clifford A. *Dinosaur.* Orange, California: Promise Publishing Co., 1987, photo page S Creation Evidences Museum, photograph by David Lines.]

tance, a human footprint was found partially obliterated by a dinosaur that had stepped on the print—either that, or it was a dinosaur foot that had human toes protruding from one side! After only two days of excavation, four human footprints had been uncovered in the same area as twenty-three dinosaur prints. They invited both geologists and the press to come and examine the finds as they were being made! Some of the later finds were filmed being uncovered and even reported in the local press.[17] (See photos below.)

Since that time, a number of additional finds have been made in the same general area including a fossilized human finger, the fossilized skeletal remains of a human being, a fossilized stegosaurus, as well as many more footprints in several layers and locations.[18] In any unbiased study these prints would be clear and incontrovertible evidence that human beings and dinosaurs existed together fairly recently. And, truth be told, there have been numerous finds throughout the world of human (or "humanlike") footprints in rock layers that are "far too old" to contain such features, though they were not found in conjunction with obvious dinosaur prints. As *Scientific American* reported in 1940, "on sites reaching from Virginia and Pennsylvania, through Kentucky, Illinois, Missouri and westward toward the Rocky Mountains, [humanlike prints] have from time to time been found on the surface of exposed rocks, and more and more keep turning up as the years go by. What made these prints? [Hmmm, what could possibly have made "humanlike" prints?] As yet the answer is unknown to science. *They look like human footprints* and it has often been said … that they really are human footprints made in the soft mud before it became rock. [However], if man made these prints in this manner, then man's antiquity is no matter of a mere million years or so, as scientists think [sic], but a quarter of a billion years, for … these rocks were laid down about 250,000,000 years ago [italics added]."[19]

Once again notice that it is the previous belief in the concept of billions of years of time that distorts and colors the far simpler and more obvious interpretation of the evidence. Who was it that said, "I wouldn't have seen it if I hadn't believed it"? A perfect illustration of the way interpretation of data is colored by prior beliefs can be seen in a *Los Angeles Times* article in December of 2005. It states that British scientists had discovered fossilized human footprints in volcanic ash that they had dated to be forty thousand years old. They consequently concluded that this "proved that humans had colonized the Americas much earlier than 11,500 years ago, as previously believed."[20] Subsequently, UC Berkley geologist Paul Renne used radioactive dating methods to analyze the rocks and came to the conclusion that the rocks in question were actually 1.3 million years old, proving that, therefore, the footprints "are not human in origin ..."[21] Incredible! No observer raised any doubt about either the authenticity or the humanity of the footprints' source until it was concluded that the prints were actually older than first believed (and this coming from the scientific community that used the same radioactive dating method to "date" the twenty-year-old lava dome of Mt. St. Helens to be 2.3 million years old!).

As the apostle Paul wrote in his letter to Timothy almost two thousand years ago, "The time will come when men will not put up with sound doctrine. Instead, to suit their own desires, they will gather around them a great number of teachers to say what their itching ears want to hear. They will turn their ears [and, evidently, their eyes and senses!] away from the truth and turn aside to myths [II Timothy 4:3-4]."

There are many places throughout the world where dinosaur footprints are found. It should be recognized that footprints preserved in rock are a very unusual phenomenon in and of themselves. Any footprints in mud, even after the mud dried and solidified, would normally be preserved only until the very next rainstorm. Even lacking rain, footprints in dirt or mud do not tend to survive in the sun for very long. Generally speaking, in order to be preserved at all, it is necessary that whatever layers these were formed in were rapidly buried themselves and left undisturbed ever since. In other words, it takes cataclysmic circumstances to preserve footprints in stone. Yet "[dinosaur] tracks are worldwide in extent. They are found in western North America and in New England. [They] are also found in South America, especially in Argentina. England also has them. And so has Basutoland, down in the southern part of Africa. In this out-of-the-way place, dinosaur tracks are quite abundant. The dinosaur hunters have also found tracks in such diverse places as Morocco, Portugal, and Australia. Canada has not been neglected either. Dinosaur footprints are also found in British Columbia."[22]

There are also features of other tyes of animal footprints preserved in stone that baffle the paleontologists and evolutionists. It is said that prints and fossils are rarely found together. Why? One source speculated that perhaps it was because rising waters floated away the dinosaurs and also buried and preserved their tracks. Could that be the result of a major flood? One researcher further noted that a particular set of dinosaur tracks started off deeply impressed, "as though the animal were unsupported by water, and became less and less well marked" as though the rising water were lifting it off the ground.[23] Many of these tracks, even though they do not show dinosaurs and humans as coeval, are nonetheless problematical for similar reasons. For example, "on one trackway ... a three-toed creature apparently took a few steps, then disappeared—as though it took off and flew."[24] That should not be a problem because the tracks appeared to be those of a bird, and flight is quite a normal procedure for such a creature. However, in the era of time supposedly represented by this particular layer of rock, birds were not supposed to have existed. One fossil track collector named Jerry MacDonald stated of this find, "We don't know of any three-toed animals in the Permian [period] ... And there aren't supposed to be any birds."[25] In another set of fossils, he found tracks of an animal walking on its hind legs. The tracks appeared to be exactly like those of a modern bear. Each footprint had five arched toe marks, like nails. An examiner commented that they look exactly like bear tracks. "'Yeah,' MacDonald says reluctantly, 'they sure do.' [But] mam-

Dating a Dinosaur: What Your Mother Never Told You

An apparent human *shoe* print found *inside* a two-inch thick slab of rock in Utah in 1968. Squashed under the heel is a trilobite that supposedly was extinct for well over two hundred million years before humans existed. The heel of the ten-inch shoe is indented in the rock and shows wear much like a modern shoe. [Photography used with permission of the Creation Research Society, P.O. Box 8263, St. Joseph, MO 64508-8263/composition by Bradley W. Anderson.]

mals evolved long after the Permian period, scientists agree, yet these tracks are clearly Permian."[26]

So, once again it would appear, the evidence is interpreted—or rejected!—based on the belief system in place.

In 1968, William J. Meister, amateur trilobite collector, split open a block of slate and found a clear human shoe imprint (see photograph). The dilemma was that, fossilized within the outlines of the shoe was a trilobite.

According to evolutionists, trilobites were among the world's first primitive organisms and became extinct over five hundred million years ago. The shale in which the imprint was found is dated from the Cambrian period and said to be between 505 and 590 million years old. The heel print of the shoe is indented one-eighth of an inch more than the rest of the sole—much like a modern shoe.[27] An evolutionary biologist from Michigan University was asked about this particular imprint. His response was, "I am not familiar with the trilobite case ... but would be greatly surprised if this isn't another case of fabrication or willful misrepresentation. There is not one case where a juxtaposition of this type has ever been confirmed. So far the fossil record is one of the best tests that evolution has occurred. I put the creationists and those that believe in a flat earth in the same category."[28] This professor had never seen the print, never examined any of the actual evidence of the case, and yet was able to pass judgment on "foolish creationists" without any reference to the available data whatsoever. Obviously, at some point, it seems self-evident that the evidence has become secondary to the theory itself. The truth is that both visual inspection and computer analysis have shown the Meister print to exactly match the size and shape of a modern shoe. The only reason for rejecting it as "human" in origin is that such acceptance would undermine both the scenario of evolution, as well as the fossil dating methods themselves.

(2) Modern Sources:

In the frozen tundra of Siberia and even parts of Alaska, a large number of woolly mammoths' remains have been discovered (see also previous chapter "The Myth of the Ice Age"). No one knows how many have actually been unearthed over the years because the natives often harvest the ivory from the tusks, and the finds have been largely unreported. One study estimated that along one six hundred-mile stretch of the Arctic coast, more than a million tons of tusks were buried. By the most conservative estimates, at least five million mammoths were buried in this portion of the world. In fact, their remains were so plentiful that one Russian scientist, Dr. Leopold von Schrenck, published a study in 1869 that concluded: "The mammoth ... is a gigantic beast which lives in the depths of the earth, where it digs for itself dark pathways, and feeds on earth ... They account for its corpse being found so fresh and well preserved on the ground that the animal is still a living one."[29] Some of the bodies found give indication of death by suffocation after being rapidly, almost instantly, buried by massive forces. Some had frozen, chewed, grasses in their mouths that had not even been swallowed. Others had undigested buttercup seeds and flower fragments in their stomachs. The variety and types of plants indicate that the environment of Siberia and Alaska was once very warm and plush. In one report mentioned by Velikovsky, an explorer in the eighteenth century found a ninety-foot-tall plum tree with fruit and leaves still attached frozen in the muck. The mammoths (many rhinoceroses and other smaller animals have been found as well) must have been overwhelmed suddenly with a rapid deep freeze and almost instant death. "The sudden death is proved by the unchewed bean pods still containing the beans that were found between its teeth, and the deep freeze is suggested by the well-preserved state of the stomach contents ..."[30]

One study concluded that in order to preserve the flesh and tissues (not to mention stomach contents!) as they were, the temperature around the animal must be dropped almost immediately to well below minus 150 degrees Fahrenheit.[31] Interestingly, in many places, the frozen soil contains many broken trees in much disorder, none of which grow today within hundreds of miles of the area. The natives call the buried trees "Noah's wood."[32] It should be noted that most evolutionists do not include mammoths in the age of dinosaurs, and these have been variously dated to the vicinity of fifty thousand years ago (up to millions of years ago, depending on who you read or believe). However, the separation of dinosaurs from mammoths is an artificial one based primarily on the realization that the mammoths in Siberia could never have remained in their current condition for sixty-five million years. In the underlying layers beneath the frozen tundra, no fossils, marine or otherwise, have been found, indicating that the catastrophe that inundated and overwhelmed the denizens of Siberia was actually the initial stages of the flood of Noah.[33] I predict that if more searches are performed in this area of the world, frozen and preserved dinosaur remains will eventually be uncovered.

In 1856 French railway workers were digging a tunnel and after setting a charge and blasting away a boulder, they went in to clear out the debris. Reportedly, they found a huge, batlike creature in the debris, which they assumed to be dead. They dragged it outside and it began to shake its wings and utter hoarse cries. They described it as shiny black with a long neck, with a beaklike mouth and sharp teeth. It died soon after and they measured its wingspan to be over ten feet long. A scientist was called to examine this creature and he determined it to be a pterosaur—something that, by current reckoning, died out millions of years ago.[34]

In Africa today, natives describe a similar but smaller creature, said to be reddish colored, with a four-to-seven-foot wingspan, as still in existence. They call it a "kongamato" and when showed pictures, picked out drawings of the pterodactyl as being most like this creature. Furthermore, current evidence suggests that some type of pterosaur still exists in Papua New Guinea where the nationals call them "ropen" or "bing pop." A pastor named Jacob Kepas is involved in attempting to document their existence, though the dense jungles and the nocturnal nature of the animals make it difficult.[35] In fact, the *Anchorage Daily News* reported in 2002 that something akin to a pterosaur had been seen by numerous people in southwest Alaska. Most reports credited the creature with a wingspan of approximately fifteen feet. As the article reported, "A giant winged creature, like something out of Jurassic Park, has reportedly been sighted several times in ... recent weeks."[36] One pilot, initially skeptical of the accounts, later saw it for himself, along with all the passengers on his plane: "He's huge, he's huge, he's really, really big. You wouldn't want to have your children out," he said.[37] He estimated the wingspan to be similar to his plane's! Nevertheless, most scientists (not having seen it for themselves) were able to quickly discount the reports, because as one said, "I'm certainly not aware of anything with a 14-foot wingspan that's been alive for the last 100,000 years ..."[38]

It may even be a possibility that something akin to T.rex remains alive in the jungles of Papua New Guinea. In fact, according to a newspaper article at the end of 1999, several people reported seeing a frightening creature in the Lake Murray area in Western Province. On December 11, 1999, several villagers canoeing saw the creature in shallow water. The following day, a pastor from a Seventh Day Adventist Church, as well as another church member, made a similar sighting. "The creature was described as having a body 'as long as a dump truck' and nearly two metres wide, with a long neck and a long slender tail. It was walking on two hind legs (this fact alone would seem to rule out any known living creature) 'as thick as coconut palm tree trunks,' and *had two smaller forearms*. The head was similar in shape to a cow's head, with large eyes and 'sharp teeth as long as fingers.' The skin was likened to that of a crocodile, and the creature had 'largish triangular scoops on the back [italics added]." [39]

Land animals are not the only ones that produce so-called anomalies. The drawings and photos of the Loch Ness Monster have produced endless controversy (seemingly matched story for story by the Lake Champlain stories from the North American continent: Champs and Nessie) because they seem to coincide nicely with fossils of plesiosaurs that have supposedly been extinct for untold millions of years. Most scientists ridicule the concept of plesiosaur being still alive, but once again, the only problem with this reconstruction is that the living animal and the fossil are separated in time (according to evolution theory) by sixty to seventy million years. However, in

Dinosaurs on the Ark

Australia, the Aborigines have legends and stories of several different dinosaur-type creatures. In western Queensland, there is a lake named Galilee that is still the source of such stories. Livestock and people in fishing boats have reportedly disappeared without a trace, and the boats have been found overturned or smashed, and "large, dark shapes have been reported moving through the dismal waters."[40]

Aboriginal drawing of "Yarru." Note the similarity in size and appearance to both plesiosaur and the Loch Ness Monster. [Illustration used with permission, <CreationOnTheWeb.com>]

This lake is not accessible like Loch Ness, however, so very few explorations have been made. The Aborigines have long ceased to live near the shore because they say a large man-eating "bunyip" lives in or near the waters. They describe the monster as being from twenty to thirty feet long with a long, serpentine neck, with bulky body and two pairs of large flippers. A former missionary was once told of a similar creature called Yarru and the tribal artist offered to draw it for him (see illustration below).

Its relative size compared to the people in the drawing and its features coincide perfectly with both the reported sightings of the Loch Ness Monster as well as the fossil record of plesiosaur.[41] These native Aborigines had

Dating a Dinosaur: What Your Mother Never Told You

never been out of the country, nor had they ever seen or heard of anything to do with Loch Ness or plesiosaur. It is absolutely beyond comprehension that they could have invented "Yarru," and had it nevertheless perfectly coincide with the fossil evidence of plesiosaur as well as eyewitness accounts of "Nessie" and "Champs."

The truth is that numerous plants and animals that were once thought to have been extinct for millions, and even hundreds of millions of years, have been rediscovered, alive and well, and completely unchanged by any evolutionary

Photograph of a living Wollemi Pine—identical to fossilized Wollemi branches. What happened to evolution for 150 million years? [Author's photograph, December 2008]

process. At the end of 1938, the coelacanth was "rediscovered" (though not identified until 1939) as a "living fossil." (A "living fossil" is a creature that is said to have disappeared from the fossil record for vast eras of time only to be refound somewhere on earth and unchanged by evolution.) In 1948, headlines in newspapers recorded that science had made a "spectacular discovery" and found a previously unknown "100,000,000-year-old race of redwoods."[42] The "dawn redwood," or *Metasequoia*, was discovered in China where it was locally well known, but only known in "scientific" circles as having been extinct for millions upon millions of years. The tuatara lizard was once thought to have been extinct for 135 million years, and yet it is still around and unchanged by any evolutionary process. In 1994, in the Outback of Australia, a hiker named David Noble discovered a grove of different-looking trees. He saved a branch and brought it back to civilization, but it

"Triassic Triops" are available online or from stores such as Wal-Mart. Notice the claim on the box that these creatures have gone "unchanged for millions of years." [Author's photo]

took two years for it to be identified as yet another "living fossil" thought to be extinct since the "age of the dinosaurs."

It is now known as the Wollemi Pine tree (or *wollemia*) and is currently being sold in seedlings as the "dinosaur tree." Also in Australia, as recently as 2002, it was reported that the "Gladiator" insect, once thought to have been extinct for millions of years, has now been rediscovered alive and well and unchanged by any evolutionary process.

On my bookshelf in my office, I have a box of "Triassic Triops," which are advertised as "actual prehistoric creatures that are identical to their prehistoric kin [see photo above]." The box also proclaims them as "unchanged for millions of years" and then pretends that this can be explained because the eggs can go into a state of "suspended animation" (*diapause*) or lie dormant for up to twenty-five years. Of course, that means that if dinosaurs became extinct sixty-five million years ago, then the dormant triops eggs would have perished 64,999,975 years ago!

I also have on my desk a "prehistoric evergreen"—"The Amazing Dinosaur Plant"—that is sold in toy stores and advertised as a plant that "once looked down to see the first

The box from the "Amazing Dinosaur Plant" that has managed to keep from evolving—or going extinct!—for almost 300 million years! [Author's photo]

dinosaurs emerge." (True, because it can still do that—on my desk, the plant overlooks my growing "Triassic Triops" that are still alive and swimming after hundreds of millions of years!) The box states, "Dinosaur Plants have lived on earth for over 290 million years [see photo above]!"

Maybe all of this evidence can be overlooked or ignored because most of the organisms are not of the type usually associated in the public's eye with "real" dinosaurs (like stegosaurus, or brontosaurus or tyrannosaurus), but it should be noted that many of these organisms—according to the evolutionists themselves—are thought to be millions of years *older* than the aforementioned dinosaurs. The likelihood of any of them still existing (if either evolution or millions of years of time were the truth) is infinitesimal compared to the possibility of T.rex still being around (which, I believe, he is).

There is, however, also current evidence that a very large, "brontosaurus-type" animal may still exist in the remote swamps and jungles of the Congo (as well as other similar areas of Africa). [Note: Most laymen are immediately familiar with the dinosaur commonly referred to as "Brontosaurus," and thus I use the name here. It should nevertheless be noted that, even though some natural history museums continue to use the name to this day, the type of dinosaur represented should be more properly referred to as "Apatosaurus." Paleontologist, Othniel Marsh, who named the Brontosaurus, or "thunder lizard," mistook a full grown Apatosaurus fossil skeleton for a new species of dinosaur and gave it the name, Brontosaurus. He also attached the wrong skull—that of a Camarasaurus—to the body of "Brontosaurus." Since Apatosaurus was named prior to Brontosaurus, when the mistake was discovered in the early twentieth-century, the first name was given precedence, and "Brontosaurus" was officially dropped. Nevertheless, the name Brontosaurus has continued to be used by laymen and scientists alike, and is not easily replaced.] In 1980, a native reported to a jungle mission that as he rounded a bend in the river in his canoe, he saw a reddish-brown animal with a

Dating a Dinosaur: What Your Mother Never Told You

snakelike head on top of a neck that was six to eight feet long and a large rounded back. Looking at artists' drawings of dinosaurs, he identified it as a type of sauropod (possibly brachiosaurus). Around the same time, a young girl in a canoe became stuck on what she thought was a sandbar. As she tried to dislodge her canoe, something kept pushing her back. "Suddenly, there was a great amount of splashing, and a large animal broke the water's surface. The terrified girl could not tell the head of the animal from its tail, but she described the body as the size of four elephants. In shock and crying for help, the girl was found by her parents sometime later. When they returned to the place where the mystery animal had been seen, they found prints of a large, unknown animal that they were able to follow for hundreds of feet along the lake shore."[43] The imprints of clawed feet and other signs in the jungle clearly indicate the existence of a very large creature that is not currently known by "science."

In fact, in 1983, Congolese biologist Marcellin Agnagna was part of an expedition to document these creatures on Lake Tele in the heart of the Congo. The previous year, on an extensive expedition with researcher Roy Mackal, they were able to retrieve convincing evidence (consistent eyewitness accounts, as well as physical habitat evidence) that an elephant-sized, long-necked, sauropod-type dinosaur still exists in the Congo. In addition, they were given eyewitness accounts of other creatures that may include the pterodactyl and even the stegosaurus. According to Agnagna's account from 1983, after being frantically hailed by one of their villager porters, "We were ... able to observe a strange animal, with a wide back, a long neck and a small head... The animal was located at about 300 metres from the edge of the lake, and we were able to adv[a]nce about 60 metres in the shallow water, placing us about 240 metres from the animal, which had become aware of our presence and was looking around as if to determine the source of the noise... The f[r]ontal part of the animal was brown, while the back part of the neck appeared black and shone in the sunlight. The animal partly submerged, and remained visible for 20 minutes with only the neck and head above the water. It then submerged completely ... [and] no further sighting of the animal took place. It can be said with certainty that the animal we saw was Mokele-mbembe, that it was quite alive, and, furthermore, that it is known to many inhabitants of the Likouala region. Its total length from head to back visible above the waterline was estimated at [twenty feet]."[44] I believe that this is a type of the animal the Bible calls *behemoth* and I also believe that God has allowed it to be preserved in small numbers in isolation until the time that its revelation will have the most impact on an unbelieving world. In an issue of the journal, *The Morning Star*, there is an article by Doug Duble entitled "God's Dinosaurs Are Coming." He describes a prophetic vision that God gave him of his plan to bring restoration to the church. At the end of his vision he reported: "Suddenly the whole throne room began to shake, and I heard a mighty rumbling, like thunder. As I looked again to the back of the hall, my jaw dropped in utter amazement at the sight before me. Dinosaurs, larger than I had ever imagined them, were proceeding toward the throne. They were all the Brontosaurus type, creatures I later learned were typically [sic] 70 feet long, 20 feet high, and 80,000 pounds of weight... 'This is My great power,' the Lord explained, 'which I am restoring to the Church. Although you do not always see it, My power is not extinct.'"[45]

I believe this prophecy could very well mean more than even the author knew. God himself represented these animals to Job (chapters forty and forty-one) as symbols of his greatest works of creation. The world has used them to create and support the lie of evolution, but God is reserving them as his trump card (along with the remains of Noah's Ark!). Notice, however, that it is to the *Church* that God's power needs to be returned and demonstrated. As it was at Pentecost, his power was first experienced by the Church, before his message was taken to the world. Evolution theory has taken root in society, not because of the success of science, but because of the failure of the Church. Christians have compromised and liberalized the Truth of God's Word.

(3) Other Historical Accounts:

The *Beowulf* poem is a single manuscript copy from approximately AD 1000. It is said to be a copy of a mid-eighth century Anglo-Saxon original that is no longer extant. However, author Bill Cooper demonstrates that its roots are clearly pre-Christian as far as the Anglo-Saxons are concerned, and the people and places mentioned in the account are clearly historical. Beowulf was not a mythical character: he was born in AD 495 and in AD 515 undertook to visit Hrothgar, the king of the Danes. Beowulf himself went on to become king in 533 and ruled in peace for fifty years, dying at the age of eighty-eight in AD 583.[46] Unfortunately, most modern translations have been subject to errors of translation that appear to be based on philosophy or mind-set, rather than linguistics. Several creatures are mentioned in the epic; one in particular is named *Grendel*. Grendel has usually been translated in modern times as "troll," even though the word for troll is known and is entirely absent from the poem. According to the epic, the monster, Grendel, bedeviled the Danes for twelve years until the coming of Beowulf in 515. By this time, Beowulf was renowned among the Danes for having killed many monstrous sea-dwelling reptilian creatures whose predatory activities were making life hazardous for the open boats of the Vikings. These creatures were called by the Saxons, *wyrmcynnes* (literally, wormkind), and *saedracan* (sea-drakes or sea-dragons). One of the king's ministers was killed by Grendel (his half-eaten head was found on the cliff-top), and as Beowulf and his men were tracking the female back to its lair, they came upon some sea-dwelling, reptilian monsters and killed several. It is noteworthy that the names of these creatures have historical roots throughout the British Isles (e.g., the Middle English word "knucker" stems from the Saxon word "nicor," which stood for a water-dwelling monster or dragon. In Sussex, a particular dragon, known as the Lyminster dragon, lived at a pool that is still called, to this day, Knucker's Hole).[47] Grendel was described as walking upright, bipedal, and "more huge than any human being." For twelve years prior to this, many Danes had attempted to kill Grendel, but his impenetrable hide had been impervious to their spears, arrows, and swords. It appears that the monster's forearms were comparatively weak and puny, and were his one weak spot.[48] (By description it could clearly be a dinosaur such as T.rex.) Beowulf was able to duck under the huge jaws, and hold the creature in an armlock while twisting: "Searing pain seized the ugly one as a gaping wound appeared in his shoulder. The sinews snapped and the [arm-] joint burst asunder."[49] This was apparently the younger and smaller of the two monsters, and when its arm was literally pulled from its socket, it fled to its lair where it bled to death. The name Grendel applied to both monsters and does not seem to be a personal name, but a known type of creature.[50] As Cooper quotes from one literature guide that appears to have understood: "In spite of allusions to the devil and abstract concepts of evil, the monsters are very tangible creatures in *Beowulf*. They have no supernatural tricks, other than exceptional strength, and they are vulnerable and mortal. The early medieval audience would have accepted these monsters as monsters, not as symbols of plague or war, for such creatures were a definite reality."[51] Jurassic Park, the *current* reality?

In truth, there are numerous accounts throughout history of humans encountering large reptilian creatures—creatures whose current existence is rejected by most scientists. However, because of the theory of evolution and the subsequent need to separate species by millions of years, these accounts are ignored or glossed over as the mythological fictions of ignorant and stupid people. The evidence of mankind's increasing mental capacity, however, is sadly lacking and quite unsupportable scientifically or historically.

Dating a Dinosaur: What Your Mother Never Told You

The word "dragon" and legends of dragons have occurred all over the world. The Chinese are said to have bred dragons, and it is recorded that the emperor employed a dragon keeper for the palace and even used dragons to pull his chariot on special occasions. Sometimes the word is used of a sea-dwelling monster, sometimes a land-dwelling creature, and other times, one that flew. In Welsh accounts, some dinosaurs in the form of flying reptiles seem to have survived as late as the beginning of this century. Elderly folk told of winged serpents that lived in the woods around Penllin Castle. Many were apparently killed because they bedeviled the poultry in surrounding farms. They were described as coiled when sleeping and they "looked as if they were covered with jewels of all sorts. Some of them had crests sparkling with all the colours of the rainbow."[52] When angry, it was said, they flew over people's heads with outspread wings. In one particular town, in the local church there is a carving of a known local giant reptile "whose features include large paddle-like flippers, a long neck and a small head."[53] There are, according to Cooper, nearly two hundred places in Britain alone where dragon activity has been variously reported.

In one chronicle from 1405, a giant reptilian creature is described: "Close to the town of Bures, near Sudbury, there has appeared, to the great hurt of the countryside, a dragon, vast in body, with a crested head, teeth like a saw, and a tail extending to an enormous length. Having slaughtered the shepherd of a flock, it devoured many sheep."[54] Apparently, archers were unable to harm it because its hide was impenetrable, so all the people gathered en mass. "When the dragon saw that he was again to be assailed with arrows, he fled into a marsh or mere and there hid himself among the long reeds, and was no more seen."[55]

In August of 1614 a strange reptile was encountered in St. Leonard's Forest in Sussex. "This [dragon] is reputed to be nine feete, or rather more, in length, and shaped almost in the form of an axletree of a cart: a quantitie of thickness in the middest, and somewhat smaller at both endes. The former part, which he shooteth forth as a necke, is supposed to be [close to four feet] long; with a white ring ... of scales about it. The scales along his back seem to be blackish, and so much as is discovered under his bellie, appeareth to be red ... it is likewise discovered to have large feete ... [The dragon flees] ... as fast as a man can run... His food is thought to be [rabbits] ... There are likewise upon either side of him discovered two great bunches so big as a large foote-ball, and (as some thinke) will in time grow to wings, but God, I hope, will ... that he shall be destroyed before he grows to fledge [spelling retained from original]."[56]

If all these are historical accounts—or have some historical roots—what happened to the dragons or dinosaurs? First of all, it should be noted that extinctions are still happening, and, indeed, the pace may be increasing. Figures vary considerably, because checking the entire globe is hardly feasible, but some scientists estimate that several species of plant, insect, and animal are becoming extinct each day. I have read figures as great as nine per hour! The fossil record and the tundra in places like Siberia indicate that the world was once a very different place, capable of supporting many more varieties of both plant and animal, as well as larger varieties of both current and extinct species. The entire world was once, apparently, a much more lush environment. Siberia was covered with trees and verdant grasses and millions, if not billions, of animals. The world's environment after the flood was not capable of supporting either the quantity or the size of the plants of the preflood world. With few animals to begin repopulation, a vastly changed environment, and much less food available, many varieties of animals simply did not proliferate after the flood as they had before. Most "dinosaurs" undoubtedly survived for many centuries in much smaller numbers, and some were eventually killed off by human beings. Others survived in small numbers until recent times, and a few may still be around.

(4) The Biblical Account:

People often ask why dinosaurs are not mentioned in the biblical accounts, as if their being missing from the text somehow validates the evolutionist's contentions. However, many other animals are not specifically mentioned anywhere in the biblical text, such as the duck, the platypus, the giraffe, the penguin, and so on. Yet all of these clearly coexist with man and there is simply no coherent reason why each specific dinosaur must be mentioned any more so than any other animal or species that is not mentioned. The Bible speaks of broad categories in creation, not individual animal kinds. In addition, even if dinosaurs were not specifically mentioned in the Bible, that omission could be no better support of their existence millions of years ago than thousands of years ago. There is certainly no separate biblical account of the independent creation of dinosaurs millions of years before mankind.

Nevertheless, I believe that dinosaurs are spoken of in the biblical account in numerous places; they just were not called dinosaurs. The word "dinosaur" was not coined until 1841 by British anatomist and superintendent of the Natural History Department of the British Museum, Sir Richard Owen; someone who, coincidentally, much to Darwin's chagrin, vehemently opposed Darwin's evolutionary theories when they were published.

In the biblical account, the dinosaurs that are mentioned are probably specific types, rather than an overall classification (the only reason the various kinds of dinosaurs are currently "lumped together" is that they seem to have two things in common: (1) they are extinct and (2) they are generally unknown in current human experience). In the Scriptures there seem to be at least three terms that could have applied to animals that are now called dinosaurs: *behemoth*, *leviathan*, and *dragon*. In addition, there are other Hebrew names for animals that have probably been mistranslated simply because no one knew what the word meant, so the translators tried to pick out a currently known animal to take its place.

The Hebrew word for dragon is *tanim*, or *taneem* (Hebrew: תנים), and it is used over twenty times in the Old Testament. Unfortunately, according to the Hebrew English Lexicon, the precise meaning is unknown. Some modern English translators seemingly did not quite know what to do with it, so it has sometimes been translated as "jackals" as well as various other known animals. For instance, in Isaiah 13:21, some English translations generally use the word "jackal," even though *tanim* is not the Hebrew word used. However, in Isaiah 13:22, the Hebrew word used is *tanim*, and yet it is again translated as "jackal." In the NIV, the word "serpent" is sometimes substituted because *tanim* could represent either a land-dwelling or a sea-dwelling dragon. Psalm 91:13 says, "You will tread upon the lion and the cobra, you will trample the great lion and the 'serpent.'"

Understanding Hebrew poetry, and its system of parallelism may help us to arrive at a better understanding of this verse. The concept is that "you will tread upon a lion, but not only that, you will trample the great lion. Then, you will tread upon the cobra, and you will trample the great dragon." The Hebrew word used here is *tanim* and it should be translated "dragon" or "great serpent" rather than generic "serpent" (which might have been one of the creatures now called "dinosaur").

Isaiah 43:20 is another passage where tanim is used, and in some versions is translated as "jackal." In the NIV it says, "The wild animals honor me, the *jackals* and the owls, because I provide water in the desert ..." The word translated "jackals" is actually tanim, which in the KJV is more accurately translated "dragons." (See also, Psalm 74:13.) However, since the context seems to be referring to real animals rather than those that may now be considered by some to be allegorical or mythological, it would appear that some translators have substituted real "jackals" in place of mythical "dragons."

In the book of Job, in chapters 40 and 41, God describes to Job the greatest animals of his creation when he talks of *behemoth* and *leviathan*. These are certainly not animals that are encountered on a regular basis today (although, I for one believe they are still around!). But, at the same time, God must be referring to animals that were within the realm of Job's knowledge and experience. If God were referencing animals that had passed from existence long before Job's time, his point would have been entirely lost on Job. (Note: some modern translations, such as the NIV, have tried to editorially pawn off these animals as the elephant and crocodile respectively. This is unlikely for several reasons, not the least of which is that there are words available in Hebrew for both those creatures and they are not "behemoth" and "leviathan." In addition, the accompanying descriptions, if taken seriously, clearly preclude both interpretations.) In the book of Job, God tells Job to "look at the behemoth, which I made *along_with you* and which feeds on grass like an ox... *His tail sways like a cedar;* ... his bones are tubes of bronze ... He ranks first among the works of God [Job 40:15-19, italics added]." This description by the creator himself of one of the greatest animals of his creation, bespeaks an enormous animal whose tail is likened to a *cedar tree*. Clearly, such a comparison could never apply to either the hippo or the elephant. Perhaps you think the hippopotamus or the elephant rank as "first among the works of God", but if you believe he also made the brontosaurus, it must at least make you wonder why he picked the hippo as his top example of creative prowess!

Job 41 speaks of the *Leviathan*, an animal that is described as awe-inspiring and incredibly powerful—"any hope of subduing him is false; the mere sight of him is overpowering [41:9]." And yet the NIV editors have the gall to suggest in the footnotes that perhaps the Creator is speaking of the crocodile! (Undoubtedly, the translators had never seen an episode of *The Crocodile Hunter* as he routinely captured crocodiles with his bare hands.) The passage goes on to say of Leviathan that "his snorting throws out flashes of light; ... Firebrands stream from his mouth; sparks of fire shoot out. Smoke pours from his nostrils as from a boiling pot over a fire of reeds. His breath sets coals ablaze and flames dart from his mouth." [Job 41:18-21] Perhaps the translators believed the Creator was capable of deception in his description of the lowly crocodile? Or that he was trying to impress Job with a "mythical" creature that never really existed? Or perhaps the Creator of the Universe is just "fudging" on the facts a bit in order to enhance his reputation with Job? Or could it be that maybe, just maybe, he is speaking the truth (according to his very nature) and the mythical accounts of human encounters with dragons are more historical than we have been led to believe?

The creation account in Genesis 1 and 2 is not unclear or ambiguous (see chapter "The Biblical Account of Creation"). Neither is the account of the Noahic Flood. It is only compromise with the secular world that has created doubt and confusion in the modern church. The list of the categories of animals created on days five and six of creation is the same as those listed in the seventh chapter of Genesis as being taken on the ark. All the air-breathing, land animals of creation—including all the dinosaurs—survived the flood on the ark, and many survived until fairly recent times. A few large dinosaurs, including behemoth and leviathan, may still be around and awaiting rerevelation. All these animals existed at the same time as man—in fact Adam named them! The biblical account is neither unclear nor confusing. It is only the philosophical manipulations of men, based on the deception of evolution theory, which some Christians have attempted to incorporate within the traditional understanding of Scripture, that has brought confusion. However, if the biblical text is Truth, then the only true "science" is that which begins with the foundational truths of Scripture. Realize also that Satan has been working overtime to blunt the impact of this evidence on the world. Most of the Church is either unaware of it, or unimpressed, or unsure what to do with it even if it were finally established as truth beyond a shadow of a doubt. The Church has been robbed of its testimony by compromise. Christians are not poised to rush into the gap with truth because we have already conceded much of it to the "scientists."

Endnotes

1 Lambert, David & the Diagram Group. *The Dinosaur Data Book.* New York: Avon Books, 1990, 289.
2 Norman, David. *Dinosaur!* New York: Macmillan, Inc., 1991, 23.
3 Jastrow, Robert. "The dinosaur massacre." *Omega Science Digest.* (March/April 1984): 23.
4 Rhodes, F.H.T., H.S. Zim, and P.R. Shaffer. *Fossils.* New York: Golden Press, 1962, 10.
5 Kroll, Paul. "The Day the Dinosuars Died." *The Plain Truth Magazine.* (January 1970): 26, cited in Morris, Henry, ed. *That Their Words May Be Used Against Them.* Green Forest, Arkansas: Master Books, Inc., 1997, 268.
6 Ibid.
7 Ibid.
8 Ibid., 268-69.
9 Ibid., 269.
10 Schweitzer, Mary, and Staedter, Tracy. "The Real Jurrasic Park." *Earth.* (June, 1997): 55, cited in Ham, Ken. *The Great Dinosaur Mystery Solved.* Green Forest, Arkansas: Master Books, Inc., 1998, 107.
11 Wieland, Carl. "Sensational Dinosaur Blood Report," *Creation.* Vol 19(4), 42-43.
12 Schweitzer, 55, cited in Ham, 109.
13 Helder, Margaret. "Fresh Dinosaur Bones Found." *Creation ex nihilo.* Vol. 14(3), (1992): 16-17.
14 Schweitzer, M.H., Wittmeyer, J.L., Horner, J.R., & Toporski, J.K. "Soft-tissue vessels and cellular preservation in Tyrannosaurus rex." *Science.* 307(5717), (2005): 1952-1955.
15 Morell, V. "Dino DNA: the hunt and the hype." *Science.* 261(5118), (1993): 160.
16 Baugh, Dr. Carl E., & Wilson, Clifford A. *Dinosaur: Scientific Evidence That Dinosaurs and Men Walked Together,* 2nd ed. Promise Publishing Co, 1991, 11.
17 Helfinstine, Robert F., & Roth, Jerry D. *Texas Tracks and Artifacts: Do Texas Fossils Indicate Coexistence of Men and Dinosaurs?* 1994.
18 cf., Baugh (*Dinosaurs*) and Helfinstine (*Texas Tracks*)
19 Ingalls, Albert G. "The Carboniferous Mystery." *Scientific American.* Vol. 162 (January 1940): 14.
20 Raksin, Alex. "Study Raises Doubts About 40,000-Year-Old 'Footprints.'" *Los Angeles Times.* (December 2005)
21 Ibid.
22 Kroll, Paul. "The Day the Dinosaurs Died." *The Plain Truth Magazine* (January 1970): 28.
23 Ibid., 29.
24 Stewart, Doug. "Petrified Footprints: A Puzzling Parade of Permian Beasts." *The Smithsonian.* Vol. 24 (July 1992): 78.
25 Ibid.
26 Ibid.

27 Cremo, Michael and Richard Thompson. *Forbidden Archeology: The Hidden History of the Human Race*. Los Angeles: Bhaktivedanta Book Publishing, Inc., First edition revised, 1996, 1998, 810.
28 Ibid., 811.
29 Brown, Walt. *In the Beginning: Compelling Evidence for Creation and the Flood*, 7th edition. Phoenix, Arizona: Center for Scientific Creation, 2001, 159.
30 Dillow, Joseph. *The Waters Above: Earth's Pre-Flood Vapor Canopy*. Chicago: Moody Press, 1981, 371-77, cited in Brown, 165.
31 Sanderson, Ivan. "Riddle of the Frozen Giants." *Saturday Evening Post*, 16 (January 1960): 82-83, cited in Brown, 165.
32 Brown, 167.
33 For additonal information and conclusions refer to the chapter, "The Myth of the Ice Age."
34 Unfred, David. *Dinosaurs and the Bible*. Lafayette, Louisianna: Huntington House Publsihers, 1990, 30.
35 Guessman, Garth. *South Bay Creation Science Newsletter*, e-mail, January 25, 2005.
36 "Giant mystery bird spotted in Alaska." *The Anchorage Daily News*. (October 17, 2002).
37 Ibid.
38 Ibid.
39 *The Independent*. Papua New Guinea (December 30, 1999), cited in *Creation ex nihilo*. Vol. 23, No. 1, (December 2000-February 2001): 56.
40 "Australia's Aborigines: did they see dinosaurs?" *Creation ex nihilo*. (December 98-February 1999): 26.
41 Ibid., 26-27.
42 Woodford, James. *The Wollemi Pine: The Incredible Discovery of a Living Fossil from the Age of the Dinosaurs*. Melbourne, Australian: Text Publishing, 2000, 64.
43 Unfred, 36-37.
44 Mackal, Dr. Roy P. *A Living Dinosaur: In Search of Mokele-Mbembe*. Leiden: E.J. Brill, 1987, 312-313.
45 Duble, Doug. *Morning Star*. Vol. 7, No. 2, 40-41.
46 Cooper, Bill. *After the Flood: The Early post-Flood History of Europe*. Chichester: New Wine Press, 1995, 146-51.
47 Ibid., 151.
48 Ibid., 154-56.
49 Ibid., 155.
50 Ibid., 156.
51 Longman Literature Guides (New York Series). *Beowulf*. 65, cited in Cooper, 160.
52 Ibid., 132.
53 Ibid.
54 Ibid., 133
55 Ibid.
56 Ibid., 134.

Chapter 12

THE SCRIPTURAL ADVENT OF ANIMAL CARNIVOROUSNESS

In addition to questions regarding the existence of dinosaurs at the time of the flood, I am often asked variations of two additional questions: (1) how could Noah have fit all those huge dinosaurs on the ark? And (2) how could Noah have fed and cared for all the carnivores on the ark?

The first question is quite simple to answer since it clearly makes sense for God to have provided Noah with juvenile animals that would take up less space, eat less food, sleep much of the time, and especially, live the bulk of their lives in the post-flood world, producing as many offspring as possible during their life span. The second question requires a bit more effort since the picture of T.rex and velociraptor—made especially popular in the *Jurassic Park* movies—as dangerous, violent carnivores is so etched in the public consciousness. And yet, that is not the picture presented in Scripture. Consequently, answering the second question will necessitate an entire chapter.

Most current Christian commentators assume that the Fall of Adam and Eve produced the advent of carnivorousness in animals. The Fall is the entryway of sin, death, and decay into the "very good" of God's creation. After the Fall, God pronounces that "thorns and thistles" will now be man's lot in life (Gen. 3:18). Thorns and thistles, therefore, were not part of the "very good" of God's initial creation and were inserted at this point (or at least, the genetic change that produced thorns was introduced). Certain animals today seem so well suited for carnivorousness that they are presumed either to have been created this way or specifically redesigned at some point. The Fall, therefore, seems to be an appropriate place for inserting any genetic changes necessary for carnivorousness. In addition, other commentators evidently feel that the Fall is the only acceptable place, other than creation week, for allowing God's direct creative acts in nature.

And yet the biblical text states nothing regarding the advent of carnivorousness at the time of the Fall. Furthermore, there is no explicit biblical reason for limiting the Creator of the Universe to creative activity only during the initial creation week or at the Fall. He certainly did creative miracles many other places within history. He changed a wooden staff into a live serpent; he created wine from water; he created leprous skin in an instant (Moses and Miriam); he often created whole skin and bodies from leprous ones; he created whole and healthy limbs (that required the creation of ligaments, muscles, etc.) where there were none; he created brand-new eyes and ears; he created fish and bread; he created gnats from dust; he brought life from death. He rested on the seventh day, not forever, and His nature as Creator was not altered by resting for a day.

In the Genesis creation account, God tells Adam and Eve, "I give you every seed-bearing plant on the face of the earth and every tree that has fruit with seed in it. They will be yours for food [Gen. 1:29]." It is clear from this account that man was neither created nor intended as a carnivore. The following verse makes it equally clear that mankind and animals were alike in this regard: God continues "And to all the beasts of the earth and all the birds of the air and all the creatures that move on the ground—everything that has the breath of life in it—I give every green plant for food."[1] God did not say, "And to some of the animals, I also give each other as food." Animals, like man, were neither created as, nor intended as, carnivores.

At the Fall, death entered the world and the earth itself was cursed: "Cursed is the ground because of you," God said to Adam, "through painful toil you will eat of it all the days of your life. It will produce thorns and thistles for you, and you will eat of the plants of the field. By the sweat of your brow you will eat your food until you return to the ground ... [Gen. 3:17–18]" Even before the Fall, we are told, God put man in the garden "to work it and take care of it [Gen. 2:15]." It seems, then, that Adam's work is not to make the earth produce, but to overcome the thorns and thistles. Consequently, there is no reason to assume from these verses that animals no longer had the vegetation to eat that God provided before the Fall.

More important, the Fall and the ensuing curse do not change man's diet. In fact, God specifically reiterates that man will continue to eat the plants of the field. Even though death has entered the world, and the earth has been cursed, and humans and animals will now face death, there is no need—physiologically or morally—for people to begin eating animals. Since humans and animals were tied together by diet in Genesis 1:29–30, I submit there is no scriptural reason for separating them here. If animals were earlier given the same dietary restrictions as mankind, and the Fall produced no change in either man's dietary requirements or allowances, there is no theological or physical reason to change the diet of animals either. In fact, there are mitigating reasons not to do so.

Prior to the flood, the world's environment was vastly different; the florae and faunae (plant and animal life respectively) existed both in vastly larger numbers and in greater sizes than are now possible,[2] plants grew in abundance in many areas that are now either wastelands or buried under ice;[3] mankind, and undoubtedly animals as well, had vastly longer life spans.[4] In this very different climate and environment, man was able to live for close to one thousand years. In fact, in looking at the biblical genealogies from Adam through Noah, it is clear that a graph of their life spans is virtually a horizontal line. There is no appreciable deterioration in life span for almost seventeen-hundred years.[5] If, as some biblical scholars and scientists contend, the dramatic changes in man's life span after the flood were due to deleterious atmospheric and environmental changes, then, according to this contention, it should be clear that the earth's environment before the flood remained constant. During this time, there remained no need or permission for man to eat animals,[6] in which case, neither is there any reason, textually or physically, to postulate the need for animal to eat animal. Indeed, several textual references make it clear that God directed such changes at the time of, and following, the flood.

One such reference prior to the flood occurs in Genesis 6:3. It is somewhat controversial because biblical scholars still debate its precise meaning. The NIV renders the verse: "Then the LORD said, 'My Spirit will not contend with man forever, for he is mortal; his days will be a hundred and twenty years.'" Some commentators understand this to mean that God is allowing the earth and mankind another 120 years before sending the flood. This, therefore, becomes the time left for Noah to build the ark.

However, there are at least two parts of the verse whose meanings remain obscure: the part rendered "contend with" and the part rendered "he is mortal." The NIV gives as an alternate rendering: "My Spirit will not remain in man forever, for he is corrupt," and then interprets the verse to mean that man's life span is what will be ultimately limited to 120 years. I believe this is the better rendering for several reasons. First, if it were God's intention to quit "contending with man" by destroying him in the flood, his plan utterly failed, for evil people sprang up rapidly after the flood. Just a few centuries later He had to scatter corrupt men by confusing their languages (at the Tower of Babel), and ultimately He came as God-Incarnate to die in our place.

Alternatively, in the light of this latter interpretation, God's pronouncement now makes perfect sense. In the pre-flood world, when people could live as evil, unchecked for almost one thousand years, they ultimately infected countless generations as well. Indeed, Genesis 6:5 makes this clear: "The LORD saw how great man's wickedness on the earth had become, and that every inclination of the

The Scriptural Advent of Animal Carnivorousness

thoughts of his heart was only evil all the time." Imagine in today's world if a man like Adolph Hitler could live for 950 years; think of how much evil he could inflict upon the world and how many generations he could sway! He could literally affect the world "forever." God's pronouncement seems clearly designed to prevent this very thing.

Furthermore, it is clearly evident on earth today that people do not live to be older than 120 years, and no amount of better health care will ever change that. With good diet, exercise, health care, and medical advances, the average age at death might increase slightly, but the life span potential will never increase. It has been established by the Creator.[7]

I once gave a talk on creation in which I proffered my view of Genesis 6:3 as God limiting the life span of man. Afterward, a man approached me and identified himself as an ophthalmologist. He had just come from a conference where the keynote speaker made the statement that, after a lifetime of study, he was convinced that the human eye was specifically designed to last a maximum of 120 years. If people could be made to live longer than that, he asserted, they would all be blind! Though I may not currently have the medical research available to support this claim, I believe that the same thing could be said for most other aspects of the human body as well. God pronounced in Genesis 6:3 that man's life span would be limited to a maximum of 120 years and current medical experience bears testimony to that fact. In fact, even if the average age of death can be made to increase because of medical breakthroughs and dietary improvements, the life span of a human will not be made to increase beyond 120 years because it has been physiologically determined by the Creator.

Some commentators have argued that, though the life span change was caused by a number of factors, it was largely an environmentally induced genetic change over time.[8] Yet, even if it could be established that changes in the environment were solely responsible for life span declines, this could not mitigate against the fact that this decline was part of God's pronounced intention. Did the One who created all life not know or understand what the aftermath of the flood would bring? Carl Wieland, creationist author and editor of *Creation* magazine, contends that because the effects of the pronouncement did not appear immediately, the "life span limitation" interpretation of Genesis 6:3 is invalid.[9] However, there are many such prophetic announcements in Scripture that find their full completion much later in time. For example, God's pronouncement to Adam and Eve that they will surely die (literally, "dying you will die") if they eat the fruit of the tree of the knowledge of good and evil, began to take effect immediately when they sinned; the effect was completed after nine hundred years.

From a completely practical standpoint, it makes perfect sense for God's plan to be established immediately, but to come to full fruition only over several centuries. Noah was five hundred years old when his sons were born (old even by pre-flood patriarchal standards).[10] If God's ultimate plan had been effected immediately, Noah's sons (who were nearing one hundred years old when the flood came) could have lived at best for a mere twenty years beyond the flood. Consequently, it would have been highly unlikely, apart from miraculous intervention, that they could have begotten any children. It would have been virtually impossible for them to repopulate the earth.

If God's prophetic announcement limiting man's life span seems slow in coming to fruition, nonetheless it is also clear that something happened immediately. Noah lived for 350 years after the flood to the pre-flood-like age of 950 years. Evidently the changed environment and diet did not affect his system. Shem, however, the first generation born after God's pronouncement, lived to be "only" six hundred years old—a reduction of nearly 40 percent. Though he was born in the pre-flood environment before the environmental changes could have affected his genetic makeup, nevertheless his life span was dramatically reduced. If these genetic changes and reduction in life spans came about solely from changes in the environment, diet, climate, amounts of harmful radiation, and so forth, then there is no physical reason why the deleterious effects should stop at 120 years.[11] Mankind's life span should, by now, have been eliminated altogether.

Ultimately, whether the reduction in life span was totally due to environmental causes, or entirely supernatural direction, it was still God's directive purpose. Either way, genetic change clearly took place beginning at the flood and continued for centuries beyond. And Genesis 6:3 announces this purpose in advance. Are we really to believe that this change was not part of God's plan, or that, worse yet, it caught Him by surprise?

Following the flood, there are two more specific references that give further clarity to the changes wrought by God. The first is found in Genesis 9:2. Here God pronounces to Noah that "the fear and dread of you will fall upon all the beasts of the earth and all the birds of the air, upon every creature that moves along the ground, and upon all the fish of the sea; they are given into your hands." If it is not fully clear from Genesis 6:3 that God initiated directive genetic changes in conjunction with the flood, it is ultimately clear that such changes did take place: mankind's life span was tremendously altered—eventually reduced by an average of over eight hundred years! In this instance, however (Gen. 9:2), it is not only clear that the "fear and dread" of man is a new pronouncement, but that it is specifically directed by God. There is simply no clear theological, interpretive, or hermeneutical reason why this change could not have included genetic change. In fact, it almost certainly would have required such change.

If Genesis 9:2 does not indicate a significant change in the relationship between man and animals, there was no reason to pronounce such change and inform Noah of it. This pronounced change also implies a general harmonious relationship between humans and animals before the flood and, by implication, among animals as well. For instance, Isaiah 6:11 states that "the wolf will lie down with the lamb, the leopard will lie down with the goat, the calf and the lion and the yearling together, and a little child will lead them."¹² Whether this particular prophecy is taken literally or only figuratively is irrelevant because the picture demonstrates God's intended idea of harmony: wolves do not eat lambs and lambs do not fear wolves; leopards and goats the same; but, also, children are not afraid of carnivores *or* vice versa. This seems to indicate that, for animals, these two fears go hand in hand: fear of each other and fear of humans. Before the flood, if animals were generally engaged in carnivorous activity, how is it possible that they were not afraid of humans at this time? Is it possible for them to live in harmony with man and yet kill and eat each other? Would they not attack and kill men as readily as any other animal? It is the fear and dread of humans that now keeps the wild animals generally isolated from mankind. Did carnivores therefore distinguish between man and other animals? We cannot argue from current experience because we now live in a world in which wild animals do fear mankind. In fact, it is said that in places like our national parks that the greatest threat to humans comes when a carnivore becomes too familiar with people and loses its natural fear.

One question that must be answered by advocates of a pre-flood origin of animal carnivorousness, is, why was the fear and dread of man made necessary after the flood if it was not necessary before the flood? Dr. Wieland addresses this question by asserting that "God's reasoning is not revealed, nor is it clear in any way that this was the basis of any 'need' ... "¹³ While it may be true that God's reasoning is not revealed, it does not follow that He had no reason, nor is there anything inappropriate in trying to discern His reasoning from His revealed Word. If God did it, He clearly did it for a specific purpose, even if the full meaning was not spelled out in detail.

It seems possible that the "fear and dread" of mankind could have been initiated by God after the flood to cause animals to disperse, as well as to protect mankind from the advent of carnivorousness. Immediately after the flood, there were only eight people surviving. Yet there were lions and tigers and bears and tyrannosauruses and many other "carnivores" all looking for food. Had these animals continued to have no fear of man, they would presumably have found people as tempting a morsel as any other potential source of food. The fear and dread of man would therefore have caused the animals to disperse and spread out more rapidly.[14] Under these new conditions of carnivorous activity beginning, the fear and dread of man would also, therefore, serve to protect animals from early extinction. In fact, this is all the more reason to see carnivorous activity as a post-flood phenomenon.

God's proclamation in Genesis 9:2-3 not only protected mankind from carnivorous animals, but aided in the dispersal of the animals after the flood. [Illustration from the author's article in *TJ: The in-depth journal of Creation*. Vol. 15(1): 2001, p. 71.]

Many commentators (secular and Christian alike) have argued against the feasibility of Noah's being able to sustain an ark full of carnivores for over a year without depleting his stock of herbivores in the process. John Woodmorappe does a commendable job of demonstrating that Noah could have provided other dietary means sufficient to sustain carnivores—perhaps even by using dried, cured meats.[15] While this possibility may be credible for the duration of the flood (though I believe it would violate God's pre-flood commands and order), I think it remains inadequate for the period after the flood (though Woodmorappe attempts an explanation by allowing for post-flood carnivores to survive on carrion[16]). If dried, cured meats were necessary to sustain carnivores on the ark, such provision would be even more necessary following the flood. During the flood, the "carnivorous" animals would have had choice herbivores all around them, which they presumably couldn't touch; after the flood, though, as the animals widely dispersed, carnivores would have found their food choices much scarcer. And if these carnivores left the ark as carnivores, then every meal could potentially have wiped another species from the earth. If the population of herbivores on the ark were sufficient to sustain the carnivores after the flood, should it not have been sufficient during the flood? No special provisions would have been required: the carnivores would simply have fed upon the herbivores with them on the ark. Even accepting that Noah may have taken as many as fourteen[17] animals of each clean kind with him on the ark does little to provide sufficient food for all the carnivores. Further assuming there were many more varieties of herbivores than carnivores does not provide much assistance because many herbivores are considered unclean[18] and were therefore only spared the flood in a single pair. The death of even *one* of these animals would have forever removed its kind from the earth. Consequently, it is not sufficient to explain how carnivores were provided for in the temporary environment of the ark; one must also account for their immediate diet upon leaving the ark. If the carnivores, who now feared man and were no longer caged by Noah, turned immediately upon the other animals, it would seem that all the effort that went into preserving such animals through the flood—and keeping them apart from the carnivores!—was for naught, because many kinds would have become immediately extinct. And yet God said that His purpose in bringing these very animals onto the ark in the first place was to "keep their various kinds alive throughout the earth [Gen. 7:3]."

The Scriptural Advent of Animal Carnivorousness

Far more tenable is the explanation that the relationship between Noah and the animals on the ark was still harmonious. There was no "fear and dread" of mankind; there were no carnivores as yet. They all ate the dried fruits, grains, berries, and vegetation brought by Noah as "every kind of food that is to be eaten [Gen. 6:21]." The animals before the flood neither feared man nor were feared by him precisely because there was no need for their separation. Therefore, I would contend that none of the animals that left the ark operated as carnivores, either on the ark, or initially upon leaving the ark. As animals dispersed over the earth, whatever specific changes God instituted after the flood took some time to take full effect in order to allow animal populations to grow and stabilize.

Another verse that has immediate bearing on the same issue is the very next one in Genesis 9: "Everything that lives and moves will be food for you. Just as I gave you the green plants, I now give you everything." Some commentators have contended that this has no bearing on the beginning of carnivorous activity either among humans or animals. For instance, Dr. Wieland suggests that "The permission given to man after the flood was exactly that—permission. Not a change in metabolism."[19] I would submit that such a position is highly untenable for several reasons. First of all, the text says nothing about "permission," either directly or by implication. God actually says, *"Just as I gave you the green plants,* I now give you everything [italics added]." In other words, "in the same manner, and for the same reason" that God gave the green plants, He now gives animals as well. He is not giving humans "permission" to eat animals; He is giving them what they need, in the same way that He once gave them green plants. Any insistence that this means only "permission" forces the unlikely conclusion that God originally gave mankind only "permission" to eat green plants, but they had no need for such food. Furthermore, left unanswered—and perhaps unanswerable—is the question, Why did God give such "permission" only after the flood? Did God decide that since immoral humans were going to kill and eat animals anyway, he might as well eliminate one sin by just giving them permission? If that were the case, why not just give them "permission" to commit adultery and numerous other sins as well? Does this not seem completely out of keeping with God's character and nature? I know of no other place in Scripture where God changed His laws, or the consequences of disobedience, as a concession to sin.

Wieland and others have concluded that immoral humans, who made choices of their own, undoubtedly killed and ate animals in the post-Fall, pre-flood world.[20] This may certainly be true, but it is a moot issue. For, even if this was a demonstrable fact rather than an a priori assumption, it demonstrates absolutely nothing about either the nature or the advent of carnivorousness. Today, shepherds keep flocks of sheep (or herds of cattle, etc., etc.) that are tamed by human contact. Periodically they may remove a sheep from the flock, kill it, and fix it for dinner. The other sheep do not begin fearing the shepherd after this. They do not attack him in return, nor do they begin eating either each other or the shepherd. Obviously, sheep are herbivores, but the point is that even if people made the immoral choice to kill animals before the flood, that, in and of itself, says nothing about whether or not animals killed each other as well. People are capable of making moral, and therefore immoral, choices; animals are not. If men killed animals before the flood, they did so on the basis of sin and not of need. The simple fact is that before the flood, both pre- and post-Fall, humans were told that they were given green plants for food, and the same was true of animals. No distinction is made between the diet of humans and animals either before or after the Fall. It is only after the *flood* that people are specifically given animals to be their food in the same way that they were once given only green plants. Consequently, since humans were given a vegetarian diet before the Fall and they had no physical or spiritual reason to change this diet until after the flood, and since animals were specifically given the same pre-Fall diet as mankind, there is no physical, spiritual, or textual reason to postulate that animals preceded humans as carnivores.

In an article in *Creation* magazine, Dr. Walter Veith made some interesting and salient points that I would

The Scriptural Advent of Animal Carnivorousness

interpret in favor of a post-flood beginning for carnivorous activity. First of all, he gave credence to my earlier assertion that the fossil record attests to the fact that we now have only a small fraction of the flora and fauna that existed before the flood.[21] This clearly implies that the abundance of the flora (plant life) of the pre-flood world made carnivorous activity unnecessary.

Veith went on to point out that "we don't know what animals ate in the past. Tooth structure is not a good indicator. The panda bear [for instance] is classified as a carnivore, but it eats bamboo."[22]

In other words, finding *Tyrannosaurus rex* and sabertoothed tigers, with sharp teeth and claws, in the fossil record says nothing about whether or not they were carnivores. Veith points out that in many observed instances, a change in environment that damages or reduces vegetation can result in carnivorous activity among animals that are normally herbivores. For example: "with the destruction of northern hemisphere forests by acid rain, ... animals like chipmunks, normally seed-eaters, will now eat animals run over on the road. New Zealand's kea parrots started to attack and eat sheep. They have the same talons and beak structure as a bird of prey but they weren't using them for this until their food source ran out. And most bears ... [only eat fish] at the time of the salmon run, because there are no berries around that early in the season. Later they become 70-80% herbivore, even though they have the 'equipment' to be carnivorous."[23]

In other words, the "tools" of carnivores are not necessarily carnivorous tools. Animals that under most conditions are now herbivores, have the same tools as those animals we consider to be normally carnivorous. It is, according to Dr. Veith, largely need that in many cases produces carnivorous activity and not nature. The implications of Dr. Veith's own analysis is that most need can clearly be attributed to the post-flood world, since that is when much of the world's flora was destroyed, never to return. Animals had no more physiological reason to begin killing and eating each other before the flood than did humans.

Dr. Veith does believe that God reengineered existing genetic information after the Fall to produce such things as thorns and thistles (and perhaps even the toxic nature of snake venom).[24] I have no argument here unless he insists that this is necessarily the *only* place where God may have engineered change. It is, as even he has noted, abundantly clear from the fossil record that major changes took place after the flood.

Another of the intriguing aspects of portions of the fossil record that gives further credence to a post-flood advent of carnivorousness, is the ratio between so-called carnivores and the herbivorous population needed to sustain them. In some places it has remained an inexplicable phenomenon that an overwhelming majority of the animals found buried in a particular region are considered to be carnivorous by nature and there are no sustaining numbers of herbivores found with them. Immanuel Velikovsky points out that, of the various fossilized animal remains unearthed at the La Brea Tar Pits in Southern California, "the vast majority of them are carnivorous, whereas in any fauna [animal population] the majority of animals would be herbivorous—otherwise the carnivores would have no victims for their daily food—requires explanation."[25] In one area, "a bed of bones was encountered in which the number of saber-tooth and wolf skulls together averaged twenty per cubic yard."[26] These animals were all clearly ensnared by a cataclysmic event because their remains are virtually never complete; they are broken in pieces and jumbled catastrophically together. Even if it were argued that these animals were swept together by the flood from over a wide region, it is highly unlikely that the flood was selective in gathering only the carnivores together and sifting out the herbivores. Indeed, it then becomes even more untenable that these carnivores existed over vast regions with no corresponding herds of herbivores to sustain them.

The Scriptural Advent of Animal Carnivorousness

However, these finds present no difficulties to the theory that carnivorousness was instituted after the flood. Large populations of saber-toothed tigers were able to peacefully coexist with wolves and other carnivores—without any herds of the requisite herbivores—precisely because they were not yet carnivorous. Recall from Dr. Veith's earlier citations that much current carnivorous activity can be attributed to environment and need. And yet it is in the aftermath of the flood that the available flora is greatly diminished in both quantity and size. Clearly, even based on the fossil record, the need for carnivorousness would have greatly increased after the flood even if we did not have scriptural indications that this is where it was introduced.

There are various "scientific" arguments that have been put forth for a pre-flood, post-Fall advent of carnivorousness (such as fossils with "kill" bite marks, coprolites—fossilized dung—with animal remains in them, etc.). I would contend that most such interpretations of the evidence are proffered by those already wedded to the belief that certain animals were always carnivores; and that dinosaurs existed millions of years ago. Even the most obvious evidence can be misconstrued by someone beginning with a false premise. For example, evidences of a global flood having caused the major geological features—and fossil record—are routinely misinterpreted because the modern belief system precludes such an interpretation. In the past, scientists were even able to put together a few human skull fragments with the jawbone of an orangutan, a bit of plaster, and a lot of imagination and proclaim it as "the best evidence available that men evolved from apes."[27]

Much speculative fiction on the character of T.rex has been published in recent years under the guise of "science." Unfortunately, few seem to recognize that it is not facts that are being reported, but interpretations. For instance, as recently as September 1999, *Scientific American* reported the results of a paper by Currie and Tanke that told of "bite marks" on T.rex skulls complete with speculative interpretations about the way T.rex fought and killed. According to the article, "these bite marks consist of gouges and punctures on the sides of the snout, on the sides and bottoms of the jaws, and occasionally on the top and back of the skull."[28] This was said to demonstrate the vicious, carnivorous nature of T.rex. And yet, three months before this article appeared, *National Geographic* recorded an interview with specialist Chris Brochu, who examined these very same wounds and "bite marks." His interpretation is entirely at odds with the "facts" reported by *Scientific American*. Brochu has talked with forensic pathologists and carefully examined the marks in question and concludes they are not bite marks: "I can't find a bite line. I've tried to match these holes with most every jaw and skull I can find, and I can't get the teeth on any skull to line up with the perforations on this jaw. So what do I think? I think it could have been some sort of infection, some periodic wounds that healed."[29] Furthermore, the finely serrated edges of T.rex's teeth show no evidence whatsoever of having chewed on meat and bones, so those who insist upon the vicious carnivorous nature of this creature simply conclude that its teeth were regularly replaced every two to three years. Yet such an interpretation is not based on the evidence itself, but entirely on the presuppositions of the claimants. The evolutionist scientist has no reason to doubt carnivorousness in the fossil record, and, indeed, much to lose if it is not there. The vast majority of "evidence" for carnivorousness in the fossil record consists of finding animals that are now considered carnivores—or animals such as T.rex with apparent design similarities to current carnivores—and declaring them as carnivores. But as Dr. Veith pointed out, that is a completely invalid approach, because many current animals that seem to have clear carnivorous features are instead entirely herbivorous. Furthermore, establishing biblically that God instituted carnivorousness in the post-flood world does not preclude the possibility that some animals scavenged from the carcasses of animals that died of other means before the flood.

Much ado has been made lately over the finds of a coprolite, suggested to come from a T.rex, that contained the bones of a smaller animal. Yet, even assuming that this

The Scriptural Advent of Animal Carnivorousness

could be proven, it does not in any way establish when it occurred, or how it occurred. In most "scientific" scenarios it had to have been sixty-five million years ago because this is when T.rex is assumed to have died out. Such an interpretation, however, presupposes the fictional nature of the biblical account of a global flood, as well as the revealed time frame that was involved. For that matter, any coprolites could have been formed centuries, even millennia, after the flood if T.rex survived the flood (which, according to Scripture, he did). In fact, such an explanation makes a coprolite survival far more conceivable. Unless a particular coprolite is part of a vast sedimentary fossil layer that is clearly a remnant of the Noahic Flood deposition, there is simply no way of eliminating the possibility that it was formed in a local event long after the flood. The biblical record should not be manipulated to encompass the so-called finds of science; the finds of science need to be evaluated in light of the revelation of Scripture.

In summary, the scriptural account establishes the created diet of both humans and land animals as herbivorous. The diet for mankind was specifically pronounced by God to remain the same after the Fall, even though man now had to toil against thorns and thistles. There is no theological or physiological reason to therefore insist upon a carnivorous diet for animals at this juncture. People had consistently long (vast!) life spans before the flood, and this was dramatically changed in the last generation immediately prior to the flood. I believe that such change was directed by God to limit the effects of evil people upon the world as well as to cope with the completely changed environment after the flood. The physiological change was proclaimed by God before the flood (Gen. 6:3), and directed to take place over several generations in order to allow the world to be repopulated. It is also after the flood that mankind is specifically given a new diet by the Creator himself. If man, who had the same diet as animals, had no reason to eat animals before the flood, then there is no scriptural or physiological reason to insist that animals had this need or nature.

In the same way that the human life span was genetically redirected, carnivorousness was instituted for both humans and animals after the flood. This process took several centuries to be fully realized, in order to allow the animal populations to become large enough and stable enough to sustain carnivores. This certainly may have involved genetic changes (whether "directed" or "allowed" it was nevertheless ultimately God's doing). John Woodmorappe recognizes that many specialized diets quite likely arose *"only since the Flood* through microevolutionary changes in the Animals—the result of variation within the created kind [italics in the original]."[30] Clearly, such "specialized" diets cannot preclude carnivorousness.

The "fear and dread of mankind" was also clearly a new feature of the post-flood world and directed by the Creator himself. It is certainly reasonable to assume that there was no "fear and dread" before the flood because there was no need. After the flood, as genetic changes began to take effect on both humans and animals, life spans began to diminish, diets began to change, and fear and dread were initiated to protect both the small population of humans as well as the surviving animals. The fear and dread would have caused the animals to disperse and allowed for populations to both grow and stabilize. With enormous environmental changes and a greatly reduced flora covering the earth, carnivorousness became part of both nature and need. Noah neither fought with, nor fed meat to, carnivores on the ark—nor did he require God's continuous intervention to prevent carnivores from killing either himself or each other—precisely because they were still herbivores.

Endnotes

1 The creatures that "have the breath of life" in them are specifically land animals—those that breathe air and dwell in the same environment as mankind. Whether this admonition would also apply to sea creatures such as sharks may remain as a topic for debate. At face value, however, since the "fear and dread" of mankind that was initiated by God after the flood (Gen. 9:2) specifically included sea creatures as well as land animals—and it was at this time that

The Scriptural Advent of Animal Carnivorousness

fish were also given to man as his food—I believe that sea creatures were also governed by this admonition until after the flood.

2 For instance, in the fossil strata have been found much larger plants than currently possible, and even a frozen fruit tree in Siberia that was ninety feet tall. Cf. Brown, Walt, Ph.D. *In the Beginning: Compelling Evidence for Creation and the Flood*, 6th ed. Phoenix, Arizona: Center for Scientific Creation, 1995, 111. Some Christian commentators have tried to maintain that the devastation of Siberia and Alaska were post-flood catastrophes, but this is untenable for numerous reasons. The devastation is continent-wide and includes frozen forests of tropical trees as well as numerous unfossilized dinosaur remains (cf. chapter "The Myth of the Ice Age"). As one example of the evidence, consider the following: "Though the ground is frozen for 1,900 feet down from the surface at Prudhoe Bay [Alaska], everywhere the oil companies drilled around this area they discovered an ancient tropical forest. It was in a frozen state, not in petrified state... There are palm trees, pine trees, and tropical foliage in great profusion. In fact, they found them lapped all over each other [1,100 feet to 1,700 feet down], just as though they had fallen in that position." Williams, Lindsey. *The Energy Non-Crisis*, 2nd ed. Kasilof, Alaska: Worth Publishing Co., 1980, 54. For much more similar evidence in Siberia and elsewhere, see also, Velikovsky, Immanuel. *Earth in Upheaval*. Garden City, New York: Doubleday & Company, Inc., 1955. The evidence clearly and consistently indicates sudden and cataclysmic destruction on a massive scale by water—the Flood of Noah and no other.

3 Fossilized tropical plants are found in numerous places now continuously buried in snow and ice. For example, many tropical plant and exotic tree fossils have been found in Greenland. Cf. Velilovsky, 43–46. Numerous tree remains are found in the frozen tundra of Siberia where now nothing of the sort will grow. Cf. Walt Brown, 111. Also, Sir Ernest Shackleton, one of the earliest Antarctic explorers, noted the presence of coal seams in the cliffs of the unexplored regions of the Antarctic. According to his observations, "This proved that in some bygone age this desolate spot had had a mild, and perhaps tropical climate, which had produced luxuriant vegetation." Worley, F. A. *Endurance*. New York: W. W. Norton & Company, 1931, 284. These coal seams were formed during and immediately following the flood and not millennia after.

4 It is possible that the very large size attained by some dinosaurs may be due in part to achieving a vast age. Cross sections of "modern" crocodiles' bones have shown structures called "Haversian canals" in some very old specimens. However, certain dinosaur bones have been found to be replete with these structures. This may imply their having lived to great age. This coincides as well with the pre-flood genealogies of men. Cf. Unfred, David. *Dinosaurs and the Bible*. Lafayette, Louisiana: Huntington House Publishers, 1990, 27.

5 Using the Hebrew text of Genesis 5, there are 1,656 years from creation to the flood.

6 Since man was said to be increasingly evil, and since we know that murder occurred at least in one instance and undoubtedly many more, it is clear that men could have killed animals as well. This has no bearing on whether or not animals killed other animals, or whether animals had the need for carnivorous appetites.

7 *Creation* magazine reader Tim Dunn points out that "normal distribution mathematics with a mean at [age] 70 years would say that 16% of us would make it to 80 years or beyond. Only about 2% would see 90, 1 in a 1000 would achieve age 100, and 1 in 25,000 to 110. The 120 year limit would be 70 years plus 5 standard deviations and very rare—only about 1 in a million living to this age." Dunn, Tim. "Letters to the Editor." *Creation* 21(1), (December 1998): 5. Dr. Wieland's response indicates that perhaps only one person in a *billion* lives to reach 120 years. If that is true, that would amount to a total of 6 people on earth today. However, there is currently no person living whose age can be verified to be 120 or more.

8 See, for instance, Wieland, Carl. "Living for 900 years." *Creation* (September–November, 1998).

9 In the *Creation* issue cited previously (21[1]; December 1998), he argues, "it seems unlikely to be referring to life spans; for one thing, many people lived far longer than 120 years, for centuries after the Flood," 5.

10 Noah alone of all the pre-flood patriarchs waited until he was five hundred years old to have children. Most had at least one child well before they were two hundred years old. I believe there are two likely reasons for Noah's delay: (1) his life's work was to build the ark. This likely took him many centuries. (2) God knew he was going to begin decreasing man's life span immediately after the flood. In order to repopulate the earth, it would be best for Noah's sons to live most of their shortened lives after the flood.

11 Furthermore, if it was only environmental and/or atmospheric changes that caused a decreased life span, Noah, who was already six hundred years old at the time of the flood should have been as affected as Shem, if not more so (because of his advanced age). The fact that he was apparently untouched by the decline indicates that the change was initiated before the flood but in the generation of Noah's sons. This would be consistent with God's pronouncement in Genesis 6:3 being a limitation of man's life span.

12 One question that has been asked of me is "Why should a wolf, leopard, or lion fear a child?" And the answer is that before the flood, they did not, because they were not carnivores. After the flood, fear was instituted precisely for this reason; to spare the children: carnivores who did not fear humans would tend to eat them–especially children!

13 Personal communication.

14 Some commentators have argued that man's superior intellect would always enable him to defeat any number of vicious carni-

The Scriptural Advent of Animal Carnivorousness

vores whatever the odds, but remember, if Noah and his sons were forced to defend themselves by killing even one lion, it would have been the end of all lions. Noah was commanded to bring two of every kind, and seven (or, more likely, fourteen) of every "clean" kind of animal as well as birds. Most interpreters would not include "carnivores" among the clean animals.

15 Woodmorappe, John. *Noah's Ark: A Feasibility Study*. El Cajon, California: Institute for Creation Research, 1996. According to his research, most current carnivores can easily survive for the necessary duration of their stay on the ark on a vegetarian diet. Others, according to him, may have utilized previously prepared, dried meats. Nevertheless, if the animals on the ark were already carnivorous and managed to survive on a vegetarian diet for the duration of the flood (over a year), many kinds of animals would seemingly have been immediately eliminated once the carnivores left the ark and began their normal feeding patterns.

16 Woodmorappe contends that carnivores exiting the ark could have subsisted for some time on carrion that was buried during the flood. I would contend that this is untenable for several reasons: (1) The ark sat on dry land on the top of Ararat for seven months while the land dried up and produced vegetation. Any exposed carrion would have decayed away long before the exodus from the ark. (2) Most buried animals became part and parcel of the fossil record rather than food for the remaining carnivores, (3) Very few current carnivores now dig through the earth looking for dead animals when there are living ones around them, and (4) Most animals buried in the flood would remain inaccessible for any carnivore looking for its next meal.

17 Genesis 7:2 states, "Take with you seven of every kind of clean animal, a male and its mate, and two of every kind of unclean animal, a male and its mate ..." The Hebrew literally reads "seven seven" clean animals. Since this is immediately followed by "a male and its mate," I believe this means seven pairs of animals. The following reference to "two of every kind of unclean animal" does not follow this construction (two two), and more likely means only one pair. Some argue (cf. Michael Kruger quoted by Ham, Ken. *The Great Dinosaur Mystery Solved*. Green Forest, Arkansas: Master Books, Inc., 1998, 128–129, or Morris, Henry. *The Genesis Record*. Grand Rapids, Michigan: Baker Book House, 1976, 191.) that this construction does not require fourteen animals but, instead, three pairs of animals with one extra to sacrifice after the flood. I would contend that seven sevens, "a male and its mate," literally means seven pairs because seven animals cannot be divided into pairs of males and females. In several places it is reiterated that *all* animals and birds are to be taken onto the ark *in pairs*, a male and its mate. Nowhere is there any indication that some animals were taken on board unmatched or singly. It is not necessary to debate the con-

struction here because if I am wrong in this regard, it only serves to strengthen my overall case by diminishing the number of animals available to carnivores as their food after the flood.

18 For a differentiation between clean and unclean animals see Lev. 11 and Deut. 14. The main requirements mentioned for "clean" animals were that they had a completely divided split hoof, and they chewed the cud.
19 Personal correspondence, Jan. 12, 2000.
20 Personal correspondence, Jan. 13, 2000.
21 His actual statement notes that "We have only a fraction of the flora and fauna that were there at first—the fossil record bears that out." (Veith, Walter. "Professing Creation." *Creation ex nihilo*, Vol. 22, No. 1, 37.) I take his "at first" comment to mean "at creation," but since he is establishing the vast difference between then and now by referring to the fossil record, it would be more accurate to say "before the flood." Clearly, he understands the fossil record to have been left *by the flood* and not by the Fall. The flood was ultimately a later consequence of the Fall, but in time they are separate events.
22 Ibid., 37–38.
23 Ibid., 38.
24 Ibid.
25 Velikovsky, 65. Interestingly enough, Velikovsky also mentions that, found in a layer of the tar pits, was a human skull—assumed to be from the so-called Ice Age—with no deviation from that of a modern Indian. It was found in layers under that of an extinct species of vulture. (cf. 66–67.) Clearly, humans existed on the American continent along with saber-toothed tigers and they were destroyed together.
26 "The Fauna of Rancho La Brea," *Memoirs of the University of California*, I, No. 2, 1911, cited in Velikovsky, 65.
27 Such was the case with Piltdown Man, which from 1911 to 1953 occupied the "best evidence" position in the public proclamations of scientists. For example, from the *New York Times* headlines of Dec. 22, 1912: "Darwin Theory Proved True; English Scientists Say the Skull Found in Sussex [Piltdown Man] Establishes Human Descent from Apes." It is said that some 500 doctoral dissertations were done on the basis of Piltdown Man alone before it was finally revealed as a fraud—made up from some human skull fragments buried and then "discovered" together with a doctored orangutan jaw. One must wonder if any of these PhDs were then rescinded (cf. chapter "The Fossil Record").
28 Erickson, Gregory M. "Breathing Life into Tyrannosaurus rex." *Scientific American* (September 1999): 46.
29 Webster, Donovan. "A Dinosaur Named Sue." *National Geographic*, Vol. 195, No. 6. (June 1999): 57.
30 Woodmorappe, 116.

Chapter 13

THE 2001 SEARCH FOR NOAH'S ARK

Travel to Turkey is always a challenge, but 2001 was particularly trying. My very first flight was prevented by severe weather from landing at New York's JFK Airport for well over an hour, and by the time we finally touched down, I had already missed my connection to Istanbul. The next plane they booked me on was also delayed—first by the nonarrival of the flight crew, then by further rain—so by the time I finally landed in Europe (now in Zurich, Switzerland) I had missed my next flight to Istanbul. I was rerouted to Geneva, but that plane was also delayed for an hour and by the time I reached Geneva, I was met on the plane and escorted by an airline agent directly to the Turkish Airline's flight to Istanbul. Unfortunately, they were not so accommodating with my luggage and it did not come at all. After spending some time at the "Lost and Found Baggage" window, I was told that my luggage would arrive the following evening—which, of course, was of no value to me because I was due to fly to Lake Van on the first plane east in the morning. After trying to make them aware of the importance of having my luggage with me on the Van flight (all my climbing gear and backpack and tent were part of the missing luggage), I had to find a hotel and wait. I was very grateful in the morning when I was lucky enough to have a bird's-eye view of the luggage being loaded onto the plane and watched my gear and suitcase being loaded. The flight to Van was on time and uneventful and I was grateful to have a ride already waiting for me so that I did not have to catch a bus or taxi. However, shortly after arriving in Dogubayazit, I realized that I had never recovered my passport from the front desk at the hotel in Istanbul. Traveling without a passport in eastern Turkey is ill-advised at best, and downright foolhardy at worst, since military checkpoints are frequent and producing a passport in order to proceed is essential. Fortunately, the military checkpoints were somewhat lax along the way, and miraculously, no one insisted upon seeing my passport. In all of my travels in Turkey, that has never happened, before or since. It eventually proved to be a bit of a challenge to track down my passport and have it shipped to Dogubayazit, but I was able to retrieve it before leaving the country.

This year, our contacts in Dogubayazit, Murat and Saim, had been able to get permission for our climb from local officials in Agri. Saim told us they had also tried to get permission from Igdir and Aralik (on the northwest and northeast sides of the mountain respectively). According to Saim, the officials compared our passports to a list of names supplied by the military of those known to be searching for Noah's Ark. Anyone on the list was summarily denied permission, even though tourists searching for nothing are readily given permission for free. Saim's analysis was that the military is very afraid of any rise of "religion" in Turkey and does not want the ark to be found. Interestingly, his analysis coincides with what Mustafa Ozturk told me two years ago. Mustafa claimed that even when/if the government gave permission for Ararat to be climbed, the military would never allow the area of the ark to be searched (he did not directly say that they knew of the ark's existence or its location, but the implications are clear—if they did not know, or strongly suspect, its whereabouts, they would have no reason to deny access to anyone, or to any particular area of the mountain). In any case, the official in Agri did not look at his list of names, so as soon as permission was signed, they hurried out of the office.

Permission, in the end, is always contingent upon military approval, however, and ours was no exception. Murat has a personal friend who offered his approval, but we were forced to wait several days (the way of life in eastern Turkey) for the final go-ahead. In the meantime, I was waiting anxiously for my passport to arrive, since, without it, I might be singled out for denial. We finally met with a local general on Monday, August 20, and were given the thumbs-up. However, just as we thought we were free to go, a soldier arrived and requested our passports. Mine still had not arrived by courier from Istanbul so I was using a photocopy and my driver's license. Saim quickly spoke up to explain to the soldier that my passport was back at the hotel and locked in a room for which we did not have the key. It was a silly and unnecessary lie. It was not even believable! All the general had to do was tell us to go back to the campground and get a key and get the passport and the lie would be exposed. Not only that, it could derail our whole effort. Thankfully, my photocopy apparently proved adequate and we were released to climb. Only after we were on our way did Murat tell us that our permission was only for a "tourist" climb to the peak from the south side. We were not officially granted access to the areas we planned to search. And, apparently, the only reason we were granted permission at all was Murat's relationship with the general.

Murat and Saim had purchased a Lada—a Russian-made Land Rover imitation (a rather poor imitation no less)—especially for our expedition. They then had a rack welded roughly onto the roof to carry our gear. I found it difficult to believe that we could actually load four full packs and four men along with a driver and even move, let alone drive up steep mountain roads normally used only for horses. Unbelievably, though, they managed to fit not only our gear but much extra for the initial foray to "high camp," and we took an additional passenger (for a total of six adults in a compact "jeep"). The drive, however, proved to be quite a challenge for the little Lada—most of our extra water was used during frequent pit stops to pour over the radiator (As I initially had seen all the gallons of water being loaded, I thought, "Finally, a trip up Ararat with plenty of water, and horses to carry it!" Most of it went to the car!). The "road" was so poor that several times we had to get out and walk to give the Lada any chance of negotiating the steep, rocky sections. In addition, the view was often disconcerting when you could see what would happen if you left the road—not to mention that the three Kurds in front would often light up cigarettes in unison and the rear windows were unvented!

We finally met up with the horses somewhere around thirty-two hundred meters in elevation (ca. 10,500 feet) and unloaded all the gear. It was a grand feeling to know that all the heaviest gear would be carried by the horses up to around fourteen thousand feet in elevation, and our packs carried by paid porters up to the ice at around sixteen thousand feet. The climb up the southern "tourist" route can only be described as interminable. It is mostly very rocky scree—and very steep—which causes you to slide back one step for every two forward; it only seems like it is two steps back for every one forward! From the point we unloaded our gear—and burdened the horses—it is a thousand vertical meters to what is called "high camp," and that is the limit of the horses' ability to carry gear. I do recall in previous years having heard Richard Bright insist several times that he would never again climb the southern route, but once we were there, it was our only option. Climbing (and climbing, and climbing, and climbing ...) caused me several times to question my sanity. My ongoing consolation was that I would not have to carry my heavy pack until I reached the glacier at about sixteen thousand feet.

When we finally arrived at high camp, we set up our tents in the dusk and the Kurds prepared tea and soup for supper. The horse handlers were then sent lower to find grazing for the horses and we were told that we would have two porters in the morning to help carry gear—even so, we all lightened our loads by leaving some gear at high camp. I never seem to sleep well when backpacking—and Ararat has been the worst of all for me. The first night at high camp I dozed off once or twice, but seemingly not for long, even as tired as I was. The situation was greatly exacerbated by a tent mate that snored loudly—I had been warned of the bears that inhabit parts of Mt. Ararat, but had not anticipated sharing my tent with one.

The 2001 Search for Noah's Ark

Though the horse handlers left, the leader, Mehmet, stayed with us, but soon began feeling quite ill. At one point, Murat felt that Mehmet was sufficiently ill that he would need to be transported down; that could very well have meant the end of our climb. At first we assumed it was mountain sickness because we had climbed from Dogubayazit to almost fourteen thousand feet in the first day with little time to acclimate. However, he was an experienced Kurdish mountain man, and it seems, in retrospect, unlikely that the altitude change would impact him and not affect someone like me who lives at sea level in Los Angeles. In any case, I asked if we could first pray for him. Richard and I laid hands on him and prayed for his healing. As we prayed, he began belching and when we finished, he seemed more at ease. When Murat asked him how he was feeling, he said he was feeling better, so the decision was made for him to spend the night. However, in the morning, even though our "porter" now claimed to be fine, it was decided that in case he had experienced altitude sickness, he should not continue the climb. That meant we were down to only one porter for the climb to the ice—another thirty-five hundred feet of vertical ascent. I must confess it was discouraging to realize that I would now have to carry my full pack of over seventy pounds the rest of the way. I do recall jokingly asking Richard how he knew it was "my" porter who had fallen ill rather than his, and he replied with a mischievous grin, "I paid."

Many experienced climbers—and most who climb in order to summit—leave the bulk of their gear at high camp and climb from there with little gear and very little weight. Weight is everything when going up. The Turkish name for Mt. Ararat is "Agri Dah," and it is said to mean "mountain of pain" or "painful mountain." Whether or not that is the true meaning, it has certainly been an apt description for me. Looking up, one can see nothing but jumbled piles of rocks going on and on without end. I don't know how many times I spied what appeared to be a ridge, only to find, hours later when I finally arrived, that there was more beyond.

Paul Thomson and Richard Bright climbing the seemingly endless boulder fields of Ararat. [Author's photograph, September 2001.]

Many, many times I found myself promising that I would never do that climb again. It took a full seven hours of vertical climbing to reach the edge of the ice at around sixteen thousand feet. That may not seem long in the course of normal events, but carrying an extra seventy pounds, wearing heavy boots, and climbing up scree and over rocks, it begins to feel as if that is all you have ever done or known! Or, worse yet, all you will ever do!

Paul Thomson (on right) and author on the way up...and up! [Author's photograph, September 2001.]

The 2001 Search for Noah's Ark

When we finally reached the ice, Richard began putting on his pack and seemed raring to go. "Of course," I mused internally, "It had to be my porter that got sick—Richard carried little weight or gear to the ice." Sour grapes, I'm sure, as Richard is the bulldog of all ark searchers and has climbed that infernal mountain more than any three other climbers combined! In any case, Paul and I were bushed and I tried to let Richard know that I could not go much further that day. At the time, Paul seemed to me to be abnormally tired, as he is much younger than the rest of us (sixteen years younger than me and twenty-seven years younger than Richard) and very fit. He had done extensive training for this climb and was in the best shape of his life. He is a very experienced mountain climber and does considerable mountaineering and ice climbing (even when there is no ark to be found!). At first, I was inwardly a little pleased that I had managed to carry a heavier pack and yet seemed to be better off than the younger, more experienced mountaineer—later we were to realize it was because things were not well with Paul.

Nevertheless, we outfitted with crampons and ice-axes and headed off across the glacier. As we traversed the glacier and circumvented the main peak to the north, I seemed to get a second wind. Crossing the glacier is inestimably easier than climbing steep and loose rock!

The author—finally smiling on the ice! [Author's photograph, September 2001.]

Eventually, we decided to make camp on the ice just below the ridge and to the northeast of the summit (about 16,500 feet). It proved to be a poor choice. Not least because it was exhausting work to dig out a trench in the ice for the tents, but because it offered no shelter from the incessant winds. I was grateful, though, that I had decided to carry my compact snow shovel along (it almost got jettisoned at high camp in order to conserve weight). It proved to be invaluable in clearing space for tents. The ice-axes were used to break up the ice, but the shovel could quickly clear away the loose ice and flatten out the surface for the tents.

By the time we set up the tents, it was clear that Paul was not well. He seemed disoriented and unresponsive. He lay in his tent with his clothes on, and we had a difficult time rousing him in order to get him into his sleeping bag. I fished out my stove and somehow in the howling wind and freezing cold, managed to boil some water for supper. I made some freeze-dried stew for Murat and Paul to share, and encouraged Murat to try and get Paul to eat some (Murat, by that time, was in his sleeping bag for the night). Next, I made beef stew for Richard and me, but before I could get into my tent to eat, Paul was throwing up. I had too much going on—camp stove burning, boiling water, cleaning up—to be of much assistance, so Richard went to his aid and prayed for him as well. Eventually, Richard and I shared the stew and that proved to be our last meal on the mountain.

That night, the wind was incessant and, at times, violent. If we had not been bodily in the tent, I am convinced it would have blown off into the gorge no matter how securely anchored it might have been. As it was, I feared for the rainfly because I knew it was not properly attached to the tent. It was anchored to the ice by a stake driven deep into the ice, but that single grommet was the only insurance we had against losing it for good. I have had some miserable nights on Mt. Ararat (the worst was in 1999 with Paul Thomson while in a storm with no tent), but this one was seemingly never-ending. It was dark, and

The 2001 Search for Noah's Ark

Richard and I were crammed into the tent along with all our gear by 7:00 p.m., exhausted and with twelve hours ahead of us. I was physically beyond tired, but I could not sleep. The noise of the wind was continuous—and not at all soothing like the sound of running water! I do not know how many times I looked at my watch during the night, but I do remember seeing 12:30, 1:00, 1:30, 2:00, 2:30, and many times in between. My one consolation was that I was not freezing and, if I put my legs on top of my pack, I could find some modicum of comfort.

When morning finally arrived, I got busy melting snow for drinking water since we were now out of water completely. Paul got up shortly thereafter but he looked none too good. Consequently, Richard and I decided that Paul and Murat would stay at the tents and we hiked over the ridge to make our way to the east above the Ahora Gorge. Eventually we roped up and I set an anchor and belayed Richard so that he could look over some cliff edges.

Richard Bright roped-up and getting ready to peer over the edge into the Ahora Gorge. [Author's photograph, September 2001.]

After doing so, we realized that we needed to go farther east to be able to find the "Ray Anderson" and "George

Stevens" sites. We remained roped together and were able to get onto a rocky outcropping right below and adjacent to the sites. The only thing we could see was ice. Not even a boulder or shadow of anything was visible. If there had been something there in the past, it has now either broken away and fallen into the gorge, or it is now covered by much more ice than previously. If the latter, the melt-back sufficient to reveal something must have to be truly significant. It would seem the tops of the mountain need be almost bare of ice. We used binoculars to scour the ice as well as the visible features in the Ahora Gorge. The Abich I Glacier seems to be almost gone and the Abich II Glacier seems too reduced to hold a structure the size of the ark.

Overlooking the Ahora Gorge and the Abich Glaciers from 16,000-foot peak. [Author's photograph, September 2001.]

After spending the better part of an hour in this area, we moved over to try and ascertain how challenging it would be to descend the eastern ridge of the gorge. Permission to ascend from that side seems to be unattainable—interestingly, that seems to be the area where Mustafa indicated to me he had "once" seen the ark. As we

The 2001 Search for Noah's Ark

Author (in rocks) at 16,000-foot peak, looking into the Ahora Gorge. [Author's photograph, September 2001.]

were looking down into the gorge from our vantage point, Richard felt that he spotted an object of interest. As he got out his camera, clouds began moving in and obscured further vision. Though he waited almost half an hour, the clouds seemed to be taking up residence for the day.

Somewhere around noon, as we were heading back to camp for lunch, we received a radio call from Paul that indicated he was too sick to stay on the mountain, and Murat was taking him back down. When we arrived at camp, their tent was down and their gear packed and they were ready to head back down to low camp. Paul seemed too weak to carry weight, so Murat managed to take both his pack and Paul's and headed down. Unfortunately, he left Paul to follow on his own, and Paul was unable to do so. As Richard and I watched, Paul would stumble and fall headlong, only to get up disoriented and begin walking in the wrong direction. We quickly realized that it was suicide to let him find his own way—one misstep and he could slide down the glacier and over the cliffs into the gorge. In his weakened and disoriented condition, he may not have had the presence of mind to self-arrest. Richard and I quickly caught up to him and half carried, half led him to the edge of the glacier and the beginning of the path of descent. We found Murat very determined that we all go down together—perhaps he had realized that he could be in serious trouble with the military if he left us alone on the mountaintop.

We discussed various options but eventually realized that, for Paul's sake, we would all have to descend. Consequently, Richard stayed with Paul and began helping him to climb down, while Murat and I once again returned to camp and took down the remaining tent and packed the packs. This time mine was overloaded since Paul had left his fuel bottles as well as all the food. In addition, I now had the tent and the rope along with all my own gear. It seemed a very long way back to the trail this time and it was around 3:30 in the afternoon when we exited the ice at the trailhead. I was praying that I would not have to carry my full pack all the way down that same day. But now, with Paul and Dick already gone, Murat and I had four packs between us. He was able to get through on the cell phone and call for assistance from below, but it would take hours before anyone could get to us. Consequently, Murat began strapping his and Richard's packs together, and finally, on top, he strapped Paul's pack. It seemed dangerous because one misstep could cause a serious fall or broken ankle, but I knew I did not have the reserves to add anything to my already overweight pack. It was embarrassing to see Murat head off down the mountain carrying three packs while I was complaining—inwardly—about my one! Even with my one pack to his three, I could not keep up with the pace (it was at least some small consolation later when I hefted the two secondary packs that Murat carried and realized that together they did not weigh as much as my one pack).

We had climbed about a third of the way down when we met two of the porters climbing to meet us. Murat and I unloaded and some of the weight was redistributed and the fresh porters took the heavier of the packs. I was given Paul's pack, and I was pleasantly surprised to find that it weighed only about half of mine (in part because he had packed extra lightly for the climb down, leaving food and fuel for us on the mountain). About halfway down to high camp, I was met by another "porter" who came for my pack. At that point I was almost reluctant to give it up after having carried it so far, but once lightened of my load, I was admittedly grateful for the relief.

Coming down can, in many ways, be worse than going up—when you slip or trip going down you tend to fall even farther down, and sometimes you become part of the avalanche you created. It is definitely harder on the knees and thighs—and each year I have lost the toenail on my right, big toe! Also, climbing down, you can sometimes see great distances ahead and it is almost painful to see the point that you are headed to, and hours later, feel no closer to it! In any case, I somehow managed to overtake Paul and Richard shortly before we reached high camp. Paul had had to stop often to both rest and to regurgitate.

Back at high camp they had tea ready for us but no food. I think I was expecting to stop there for the night (it was now after 6:00 p.m.) and fix supper. I had eaten a candy bar earlier in the day, but that was the only sustenance other than water that I had managed. Before coming down, I was able to fill one water bottle at the edge of the glacier, but I had "lost" it in the switching of packs earlier. At that point, my legs were tired, I was physically tired as well as exhausted from lack of sleep, and even though I was not looking forward to setting up a tent, I was grateful that the day's journey was ended. Little did I know!

Somewhere around that time, Murat announced that the "jeep" was coming to low camp to pick up Paul to take him to the hospital. That meant that we still had to get him down that same evening. The horse handlers were bringing up the horses to get our packs and gear, so Murat and one other Kurd left with Paul to get him down to low camp. Murat wanted Richard and me to head down on our own and meet them at base camp. I'm not sure I was thinking too clearly at that point or paying attention too well, or I would have been more stunned. It had taken us close to seven hours to climb from low camp to high camp. Going down without packs would make for a much easier climb, but it would still mean that we would end up climbing at night. I must confess that I did not even think about how far we had to go or I might have simply refused. In any case, Richard and I changed into tennis shoes, dropped our packs, and foolishly headed down the mountain without taking any water or even a flashlight. It was only when it began to darken about 7:30 p.m. that we began to realize we might be in trouble. We were only about one-fourth of the way down (though neither of us fully realized it at the time) and it was getting very difficult to follow the trail. At one point, we differed as to which side of a ridge the trail followed. Richard was convinced he saw horse hoofprints to the left, but I was convinced by what appeared to be a trail marker to the right. My reasoning went that horses may wander several places, but only people build trail markers. When Richard insisted upon going his way I followed for some distance. It did not take long, however, before it became clear that there was no trail in that direction that a horse could follow. Consequently, we had to climb back up to where we had lost the trail and hope that the horses that were following would come our way.

The night was now completely black, without even a sliver of moon showing, and even knowing where the proper trail was would not have helped. Providentially, the horses and their handlers came to where we were waiting before deciding they could not follow the trail from that location. It was too steep and dangerous to take the horses down that way by night, so they had to take a long detour up and around the rocks to get back to the trail. They pointed us in the right direction and indicated we could climb down the ridge and meet them on the trail farther down. As we tried to negotiate our way down the steep ridge, without a trail, in the middle of the night, Mehmet (the man we had prayed for just two nights earlier) somehow found us and began to guide us to the trail. Without him we may have found the trail, but we could never have followed it.

The 2001 Search for Noah's Ark

Richard and I slowly (and painfully) followed our new guide down the "trail." It is hard to describe what it is like coming down a steep, rocky, and often treacherous mountain trail in utter darkness without so much as a match for a light—especially when already at the point of utter exhaustion. If it were not for my ski poles and the Grace of God, I am fully convinced I would never have made it. After we had been stumbling along—literally—for what seemed like hours (it was actually only an hour and a half after Mehmet "rescued" us), we came to a man waiting with a horse "for Richard." I have never been so jealous. But knowing that he is twelve years my senior, and after seeing him fall several times, I was grateful for the provision. I must confess that several times Richard offered to trade places with me. It was only my pride and ignorance that prevented me. Pride because I did not want to admit that I needed a ride, but ignorance because I had no idea, in the dark, how far we still had to go. I assumed that if we were in an area that it was safe for Richard to ride, we must be very near the low camp. In retrospect, if I had known we still had almost two hours of descent before us, I would gladly have accepted a ride for a portion of it.

At some point we could see a light below us that betokened camp and safety, but it never seemed to get any closer. At times I could have sworn that they kept moving it farther down the mountain—like the proverbial carrot in front of the mule, it was there to entice us but not to satisfy. Nevertheless, it served to keep me going. Miserably, though, when we finally arrived at around 10:00 p.m., it was only to walk by and keep going! I only found out the next day that the "light" came from the camp of a group of international climbers. Paul and Murat had stopped on their way down for a cup of tea, but we were in a hurry to get to the "jeep." I don't think I have ever been so discouraged in my life as the moment we bypassed the camp and I realized there was not even a cup of water in it for me. The Kurds, it seems, go all day without a drop of water—it never seems to occur to them to actually carry it with them. I've never met a Kurd on the mountain—with or without horses—that actually had water with him. Yet they usually seem grateful if I offer them some. It's a strange culture that uses cell phones on the mountain, but never bothers to carry a flashlight even while negotiating difficult mountain terrain in the middle of the night. In any case, Richard and I had set off with no pack and no water—and no place to carry it because both hands were required for trekking poles!

I don't know how many times I fell in the night. Falling completely was a rarity because of the balance afforded by the poles, but I twisted an ankle, stubbed a toe, or stumbled over unseen rocks too many times to count. I do, however, recall one instance when we had temporarily reached a fairly level, broad stretch of path when I began to walk with some confidence—always swinging my poles ahead of me at shoulder width in order to ward off the unseen rocks—and tumbling head over heels over a basketball-sized boulder directly in front of my feet. I think my Kurdish guide was painfully befuddled over how slow I was as he seemed able to walk at a normal pace in utter darkness with nary a stumble. I think I actually felt some sort of perverse satisfaction when my guide stumbled once and nearly fell.

Finally, I spied the lights of a vehicle some distance ahead, but just as it seemed to be approaching, the trail turned away and moved in the opposite direction again. I think, at that point, I actually began to wonder if I could trust my guide to take me to Murat and the jeep. Richard and the horse had long since disappeared into the night, so I was on my own trying to follow a Kurd who spoke no English. In the darkness I could not tell that the trail was circumventing a gully between us and the jeep. It was not until 11:00 p.m. that we finally arrived to find the jeep almost loaded and ready to go.

I'm not sure how long it took to finish loading, but I did manage to find a few drops of some weakly lemon-flavored water ("su"). I have no idea where the water originated, but at that point my lips were so parched, and my throat so dry, that the memory of contracting giardia from untreated water the previous year never even surfaced.

The 2001 Search for Noah's Ark

It was immensely gratifying but not nearly enough. My throat actually remained parched until the following day.

The drive down also proved to be an adventure because the Lada had sustained damage on the way up. Driving down at night, I was actually grateful that I could not see over the edge into some of the ravines. For some of the way, two men would walk or run ahead of the jeep, moving rocks or pointing the direction to go in order to avoid the worst pitfalls. By the time we reached the highway it was nearly 1:00 a.m. and the car seemed to be barely running at times. It would not accelerate and could not sustain highway speeds. I was praying fervently that we would not have to walk anywhere farther that night. We managed to get Paul to the local hospital around 1:00 a.m. only to have the doctor give him a cursory glance and release him with a prescription for stomach medicine. In the morning, he had improved to the point that he was able to eat some food, but, even though he did not throw up again, it was several days before his stomach pains had subsided. His condition prevented us from considering another attempt to get to the eastern ridge.

The rest of my stay in eastern Turkey had little to do with searching for Noah's Ark, but we did manage to make it to the village of Korhan for a day. Korhan is said to be a "very ancient" settlement and there are the ruins of a very old village there. Some have said that the "House of Shem" (son of Noah) is there still. I cannot say that I saw his house—or his burial place—but we did manage to climb one hill overlooking the valley where there are the ruins of an old altar. Paul, when he was the prisoner of the PKK (Kurdish terrorists) in 1993, was told by his captors that this was the "altar of Noah." The military who occupied foxholes on the hill at first tried to wave us away, but I pretended not to understand their motionings, waved back, and simply continued climbing. When we reached the top and they realized we were harmless American tourists, they allowed us to stay and take photographs. The altar was interesting and had apparently been used for burnt offerings, but if it had anything to do with Noah, it must have been rebuilt several times in the interim.

"Noah's Altar" on hilltop near village of Korhan. [Author's photograph, September 2001.]

On the way back to Istanbul, I also managed to spend a day at Lake Van by taking the ferry to Akdamar Island where there are the remains of a church built around AD 900. The water of Lake Van is quite salty and it is an enormous lake that would seem to be left from the receding waters of Noah's Flood. Interestingly, many people have claimed to see a "sea monster" in the lake that is reminiscent of the Loch Ness Monster. I have actually seen a video of the monster that is quite interesting, but the video monster does not resemble the plesiosaur-type creature of Loch Ness.

Knowing where to go (and when) from here is going to have to be a matter of much prayer. It seems clear that the military will never "allow" a peaceful exposure of Noah's Ark, and it seems unlikely that the various organizations vying for permission will ever get it in the way that they hope. Monitoring the glacial melt-back via satellite data is one option but would require either considerable funds or a willing source. Negotiating with Mustafa Ozturk to take me to where he claims he once saw the ark is another option, but at this point he is in exile (from the law) in Istanbul, so that would require some miraculous intervention as well.

Chapter 14

ARCHAEOLOGICAL ANOMALIES

One of the questions that is often asked of those with a "young-earth" persuasion is, if men and dinosaurs did coexist, should there not be more evidence of civilization buried in the fossil record along with the dinosaurs? And the answer is, "Yes and no." There have been numerous finds that fit neither the evolutionary scheme, nor the standard "millions-of-years-of-time" dating methods (see also chapter "Dating a Dinosaur"). However, most often such finds are of a nature they can be simply labeled as "anomalies" and then ignored.

An anomaly by definition is something that should not be: it is inconsistent with what is known or believed to be true. But how is it possible that archaeological "anomalies" exist at all? Are not the artifacts unearthed by archaeologists and/or geologists the very jigsaw puzzle pieces that are utilized to create the picture of the jigsaw puzzle? How can the pieces of the puzzle not fit the picture? They are the very pieces that should be used to *create* the picture. If the pieces of the puzzle do not fit our perception, it must be our perception, or preconceptions, that are in error, and not the artifacts themselves. But, as with the evidence of human and dinosaur coexistence seen in a previous chapter, the only real "anomaly" in the artifacts from archaeology is the belief or perception that the world is billions of years old, and that humans are fairly recent in their place in time on earth. The reality is that the Flood of Noah was sent precisely in order to destroy the civilizations of earth and the evidence of their existence should be found, not only scattered throughout the fossil record (as it is), but underneath the layers of sedimentary rock in which the fossils are found. And the finds of archaeology actually fit quite well with this reality.

Charles Lyell attempted to sarcastically denigrate the idea that advanced civilization(s) had previously existed and had somehow disappeared or been catastrophically destroyed. His comments were intended to show that, in looking at archaeological finds from sedimentary layers, there was only limited evidence of very "primitive" humans. As he put it, "Instead of the rudest pottery of flint tools, so irregular in form as to cause the unpracticed eye to doubt whether they afford unmistakable evidence of design, should we not be finding sculptured forms surpassing in beauty the masterpieces of Phidias or Praxiteles; lines of buried railway and electric telegraphs, ... astronomical instruments and microscopes ... and other indications of perfection in the arts and sciences ... ? Vainly should we be straining our imagination to guess the possible uses and meanings of such [past inventions]—machines, perhaps for navigating the air or exploring the depths of the oceans or calculating arithmetical problems beyond the wants or conceptions of living mathematicians."[1]

Lyell may have intended his comments as obvious rejoinders to those who might believe that an advanced civilization (or system of civilizations) previously existed, but Lyell overlooked, misunderstood, or was unaware of, the existence of just such evidence. Many of the items presented in this chapter came to light only after Lyell's time, but some of the things that did exist in his lifetime were overlooked as "anomalies," or misunderstood. The truth is that there is a considerable body of evidence that points to the existence of advanced civilizations prior to the Flood of Noah (see also chapter "The Pre-Flood Pyramids").

Occasionally, the ramifications of such evidences have been so unacceptable to secular "historians" that the evidence itself has been covered up, or worse, destroyed. At numerous points in history, during various wars or "purges" by despots, many ancient documents—sometimes entire libraries—pointing to our past knowledge and achievements have been destroyed. Hints of their content still remain, however, and "anomalies" in the fossil record actually abound.

Archaeological Anomalies

In 1928, Atlas Almon Mathis was reportedly working in a coal mine in Oklahoma about two miles north of a town called Heavener. It was a shaft mine that was reputedly almost two miles deep. The miners had to descend in elevators and fresh air had to be pumped down to them. One day after using explosives to blast coal loose, they found several concrete blocks lying in the cavern. According to Mathis's account, the blocks were twelve-inch cubes, polished so smooth that each side could be used as a mirror. When broken open they appeared to be made of concrete or some similar man-made substance on the inside. At one point in their ongoing excavation, part of the coal seam collapsed and the miners found a solid wall made of these cubes inside the coal bed. Over one hundred yards farther down the shaft, they found another portion of the same wall. The coal was said to be from the Carboniferous period and therefore at least 286 million years old. So, it would seem, either there were highly intelligent men existing around three hundred million years ago—long before most of the dinosaurs!—or the entire dating system is erroneous. Neither possibility bodes well for the fiction of evolution. According to Mathis, other miners also told of finding a solid block of silver in the shape of a barrel with the prints of staves on it. Shortly after the block wall incident, the company officers pulled the men out of the mine and forbade them to speak of what they had seen. The mine was subsequently closed.[2]

Perhaps if this were one apocryphal story it could be set aside as "anomaly" until there was better supporting evidence than the hearsay of one presumed "eyewitness." However, there are too many such stories to ignore, and many of them have retained the evidence. For example, in 1868 in Ohio, James Parsons claimed to have found a large, smooth wall with several lines of what appeared to be hieroglyphics in the midst of a coal mine. In 1912, Frank Kenwood found a chunk of coal too large to use so he broke it open with a sledgehammer and an iron pot fell from the center (see photo below). It was kept for many years in a small museum, but was "lost" when the museum owner died and the property scattered. In 1891, Mrs. S. W. Culp broke open a piece of coal and found embedded in it a small gold chain about ten inches long of what was said to be "antique and quaint" workmanship.³

Small iron pot found inside a chunk of coal in 1912. [Photograph from Hovind, Kent. "Creation Seminar Slides," CD. Pensacola, Florida: Creation Science Evangelism.]

In an article that appeared in *Scientific American* (from an era when they were more likely to publish such things—and during a time when Lyell could certainly have been aware of it), it was reported that in 1852 an explosion—the source of which is not clarified—threw large pieces of rock from the earth in Dorchester, Massachusetts. Inside one of the rocks was a "bell-shaped vessel" that had been broken apart when the rock burst. The vessel was described as being "zinc in color, or a composition metal, in which there is a considerable portion of silver. On the side, there are six figures or a flower, or bouquet, beautifully inlaid with pure silver, and around the lower part of the vessel a vine, or wreath, also inlaid with silver [see photo below]."⁴

Ornate "silver" vessel from "600-million-year-old" rock. Apparently those trilobites were remarkably clever with their pincers. [Ibid.]

According to the article, there is absolutely no question that the "bell" had been inside a rock deep within the earth—rock that was subsequently "dated" to be six hundred million years old. Ironically, the writer of the article remarked that the vessel must have been made by "Tubal-Cain," thereby insinuating that it could only have been a preflood relic.[5]

In 1899, in Nampa, Idaho, a well-digging drill brought up a small, handmade, clay figurine, from three hundred feet below the surface. The stratum was said to be two million years old (note: it is now reputed to be twelve million years old!). Professor F. W. Putnam noted that there were iron incrustations on the surface, which could not have been fraudulently attached, and which verified it as being of great antiquity.[6] Apparently, even though no humans existed at the time (on the evolutionary scale), there must have been some very capable apes that were prophetic in their ability to predict what future creatures would eventually look like (see photo below).

Two-million-year-old clay figurine (or twelve million!) found in Nampa, Idaho. Perhaps it was made by primitive but prophetic apes? [Ibid.]

One of the most damaging "anomalies" (to the theory of vast eras of time and evolution) was uncovered in 1922 by John T. Reid, a mining engineer and geologist in Nevada. He found, imprinted in vastly old rock, what appeared to be the interior portion of the latter two-thirds of a shoe (see photo below). The heel portion appeared to be slightly worn just as a modern shoe would wear with use. Around the outline ran a well-defined, sewn thread at regular intervals that attached the welt to the sole. In fact, geology experts agreed that the rock had come from a Triassic formation, and shoe manufacturers agreed that it was a hand-welted, human shoe-sole. By any conventional rendering of "history," the two options are entirely incompatible. Reid took the fossil to a microphotographer of the Rockefeller Institute who enlarged the photos twenty times their original size. These enlarged photos clearly showed even the minutest detail of the thread twist and warp that was used to sew the sole. Interestingly enough, H. F. Osborn, who was responsible for inventing Hesperopithecus (otherwise known as "Nebraska Man") from an extinct pig's tooth (see chapter "The Fossil Record"), reported that this fossil was not really a human shoe imprint. It was, instead, according to Osborn, "the most remarkable natural imitation of an artificial object [he] had ever seen."[7]

Partial sole of a shoe found in Triassic rock that is supposedly 213-248 million years old. [Photograph from Cremo, Michael and Thompson, Richard. *Forbidden Archeology: The Hidden History of the Human Race*. Rev. ed. Los Angeles, California: Bhaktivedanta Book Publishing, Inc., 1996, 1998, p. 807.]

Such men are spoken of in II Thessalonians 2:10–12: "They perish because they refused to love the truth and so be saved. For this reason God sends them a powerful delusion so that they will believe the lie and so that all will be condemned who have not believed the truth ..."

In *The Hidden History of the Human Race*, authors Cremo and Thomson tell of numerous human artifacts and even human remains being discovered in the early days of

Dinosaurs on the Ark

gold mining in places like California. Most such reports are now merely anecdotal because very little effort was taken at the time to properly document or preserve such finds. The discoveries were interesting, but not earth-shattering, largely because Darwin's theory of evolution had not yet taken root in society. Few people saw the discoveries as anything more than of passing interest.[8] However, some finds have been documented and even maintained. In June of 1936, near London, Texas, near the Llano River, Mr. and Mrs. Max Hahn found an unusual rock. Jutting from one side of the rock was what appeared to be a piece of wood, or a wooden dowel. The rock was originally identified to be Ordovician, which would have made it hundreds of millions of years old by conventional dating methods—before trees are said to have come into existence. Analyses now vary and it has been dated as recent as 135 million years old.[9] It was not until 1946 or '47 that the rock was broken open by the Hahns' son, George. Inside the rock was discovered a hammer with a rectangular-shaped head (see photo below).[10]

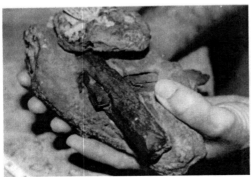

Top: photo of rock with wooden protrusion. Bottom: broken rock exposing hammer inside. [Baugh and Wilson. Photo page Q. Creation Evidences Museum, photograph by David Lines.]

The head of the hammer was tested at Battelle Labs in Columbus, Ohio, and discovered to be 96.6 percent iron, but with 2.6 percent chlorine. Metallurgists have concluded that an alloy of iron containing 2.6 percent chlorine could not be produced in today's atmosphere. Other analyses have preliminarily indicated that the head of the hammer is of very good quality and is evidence of high-technology metallurgy.[11]

Authors Cremo and Thomson, in *Hidden History*, speculate that such finds could indicate the existence of humans hundreds of millions of years ago. Neither author ever seems to even wonder about the accuracy of dating methods—even those that have led to highly discordant claims. The simple truth is that such finds point to the civilization that existed before the flood (with a significantly different atmosphere in existence). The hammer and all such other "anomalous" artifacts point directly to the accuracy and validity of the biblical account of history and the global flood that occurred about forty-five hundred years ago.

In 1971, it was reported that a hiker and amateur archaeologist found the fossilized remains of human beings in layers of rock that had been dated to be one hundred million years old. "Dr. Marwitt [J. P., professor of anthropology, Utah University] pronounced the discovery 'highly interesting and unusual' for several reasons. As the bones were uncovered, it soon became apparent that they were 'in place' and had not washed in or fallen down from higher strata... The rock and soil that had been above the remains had been completely covered by about fifteen (15) feet of material, including five or six feet of rock."[12]

Naturally, the writer of the article reporting the incredible find concludes that, since it is "known" that humans are a recent occurrence, the evidence cannot really be taken at face value: "Despite evidence that these human remains are 'in place' in a formation 100 million years old, the probability is very low that they are actually that old. The bones appear to be relatively modern in configuration, that is, of *Homo sapien* rather than one of his ancient, semi-animal predecessors [italics in original]."[13] Actually, it should be "reported" that the probability of the rock (in which the remains were found) being one hundred million years old is … nil. In fact, the remains can be reliably dated to approximately forty-five hundred years ago—they were two people who missed the ark and were caught out in the rain and mud! They are currently real rock stars!

Controversy was still being expressed over these particular bones several years after their initial discovery, and in 1975 it was again reported that the bones were unquestionably found "in place." In fact, the bodies were not broken apart, indicating they had been buried intact. Dark organic stains were found in the rocks around the bodies indicating that the bodies were inside the rocks when the rocks were formed. Barnes, who in 1971 had speculated that the bones could not possibly be as old as the rocks they were found in, concluded in 1975 from the weight of accumulated evidence that "it is highly probable that the bones are, indeed, this old, … [and] we may never know exactly how human bones came to be in place in rock formations more than 100 million years old."[14]

He may never "know," but it is no mystery when the biblical account of the Noahic Flood is acknowledged as reality. Apparently, the university scientists never attempted to independently "date" the bones (apart from their determination of the age of the rocks themselves), and, not being able to explain away the dilemma, they quit trying, and transferred the bones to a location out of sight and out of mind. Which is exactly what happens to most "anomalies."

It has been reported that in Naples, Italy, in 1503, during a huge storm, a "section of mountain split open revealing the remains of a ship enclosed within the rocks of the mountain. (Clearly there were ships other than Noah's Ark before the flood, but they did not survive the catastrophic nature of the flood.) The remains of the ship were observed by many people and it was noted that the vessel was much larger than the medieval ships sailing the seas at that time. The wood of the ship was petrified, indicating that it had been enclosed in the rock for thousands of years."[15] The remains of other ships have also been reported buried in the rocks of the earth (perhaps the dinosaurs were more capable than we have previously conjectured!). In Bern, Switzerland, for example, it was reported by miners in 1460 that the remains of a wooden ship, complete with an iron anchor, were found in the interior of the mine over one hundred feet deep in the earth.[16] And in 1540, in Peru, while digging a shaft looking for silver, miners also reputedly encountered the fossilized remains of a large ship.[17]

It should be clear that "anomalies" abound in the fossil record, and yet many still question why, if humans really existed from the beginning of creation, the evidence of human occupation is not more prevalent in the fossil record. The answer, I believe, is twofold: (1) *cities* are generally well attached to the earth and do not float in waterborne sediment, and (2) water has a tremendous filtering and sifting capacity. For instance, take a few different grades of sediment or sand, put them in a jar with water, shake it up, and watch it settle—it invariably settles into layers. Evolutionists have pointed to the layers containing different types of organisms as evidence of different eras of time (but that is falsified by the so-called anomalies that exist throughout the entire record). Yet two things should be readily apparent even in today's world: (A) First of all, different types of organisms exist in different regions and ecological zones. A flood today, for instance, could hardly capture a kangaroo and a polar bear together (unless it inundated the local zoo). (B) Second, humans, for the most part, live in concentrated locales that are quite distinct from the animals. In general, people would not be buried together with the animals precisely because humans do not live with the animals. Furthermore, all of the animals entombed in the rock layers must have been "floating" within the waterborne sediment in order to be fossilized at all. Not only would the different densities and sizes of the animals cause them to be generally deposited in different grades of sediment, but most human artifacts, like hammers, or gold chains, or iron pots and pans, do not float! In fact, cities are anchored to the ground precisely to keep them from being blown or washed away (and the vast majority of people would have been "anchored" in their homes or other buildings if rains and floods began happening. It would have been the rare individual indeed who was out hiking around with the dinosaurs when the floods came). As mentioned in the chapter dealing with the great pyramids of Egypt, floods do not create the same kind of destruction everywhere. In the areas of the earth where the "fountains of the deep" erupted, huge amounts of sediment would have been created, and any cities in its path would have been buried at the *bottom* of the layers. The fossil record—where it exists—would necessarily lie on top of any evidence of civilizations. Precisely where no one is looking!

However, there would also be places on earth where the floodwaters rose and receded without bringing vast new layers of sediment. And in fact, that must be the case, for not all areas of the earth are covered by sedimentary rock layers replete with fossils. Consequently, it seems likely that at least some of the world's anachronistic architectural features—like the Great Pyramids—are actually remnants of the preflood world.

There are indeed hints at the civilizations that existed before the flood in numerous cultures and artifacts throughout the world. As Charles Berlitz admits (while denying the possibility of Noah's Flood), "The scientific knowledge of matter and the cosmos extant in some parts of the ancient world seems to have been present on a high level *almost at the beginning* and then to have slid into a decline as if something had happened to cause civilization and knowledge to run backward. *It was as if a great catastrophe occurred* and isolated groups on the world's surface had retained bits of an advanced culture and had then gradually lost them after being cut off from the source [italics added]."[18] Yet most cultural, historical allusions to the preflood world are dismissed as myth and nonsense. However, as Berlitz is forced to admit, "there are concrete references that have subsisted for thousands of years, indicating an acceptance of pre-flood civilization by ancient cultures. King Ashurbanipal of Assyria was referring to relics from a civilization that to him was already 'prehistoric' when he had written for posterity:—'I understand the enigmatic words in the stone carvings from the days before the Deluge.'"[19] King Ashurbanipal is not a mythical character, and he is clearly referring to real artifacts that he had seen and attributed to the preflood world.

In fact, in several places in the world (besides the Great Pyramids of Egypt), there exist the remains of inexplicable architectural works whose real history seems to have been lost even by those who live there. In the Egyptians' annals and hieroglyphics, there remains no account of the construction of the pyramids—inexplicable, if indeed they were built by the Egyptians. The Romans used a preexisting stone foundation, the source of which they never discovered, as a base for building the Temple of Jupiter. The stones of the ancient foundation weigh up to four million pounds each and were precisely fitted together in a way that the Romans could not duplicate. In Sacsahuaman, Peru, there are foundations built of gigantic, and perfectly fitted, boulders that the Incas utilized as foundations for their own buildings, but whose origins were attributed by the Incas themselves to "the gods." In Bolivia, in the city of Tiahuanaco, there are enormous ruins, the builders of which are lost to history, and pottery shards that apparently depict various dinosaurs. And even though it is generally acknowledged that Tiahuanaco, with its stone docks, must have originally been a seaport, it is now at an altitude of 13,500 feet, as though it rose with the mountains.[20] In addition, in the Loltun Caves of the Yucatan, a giant stone jaguar exists that is very different from other art forms of the region. A thorough examination revealed "petrified sea fauna in its pitted carvings, demonstrating that it had been long immersed below sea level and had surfaced again as it followed the heavings of the land into and out of the sea."[21]

Clearly the occurrence of a global flood becomes an essential tool to the proper interpretation of both geological and archaeological evidences. Indeed, without that foreknowledge, proper geology and archaeology cannot be and have not been done.

In addition to the anomalies in the fossil record, and the inexplicable architectural structures remaining on the earth's surface, it is becoming increasingly clear that much evidence for a preflood civilization lies buried at the bottom of the world's oceans. There is nothing unexpected in this based on the biblical account of a global flood. It would seem apparent that much of the water now filling the ocean basins was previously contained under the earth's crust, and consequently, many areas of preflood civilization are now submerged. As Berlitz admits, even while attempting to rule out a global flood, "While there are submerged prehistoric ruins in other parts of the world, still standing giant walls and step pyramids one-and-a-half miles deep in the Peru-Nazca trench, unidentified undersea ruins off the islands of the South Pacific, Japan, and Southern India, it is probable that all these architectural vestiges are connected with the last great planetary catastrophe, which has become a part of world folklore through the memory of Atlantis."[22] Atlantis, but not the Flood of Noah?

The reality is that the biblical account of a global flood is not folklore and myth, but the true history of the earth. There are no archaeological "anomalies" because, with a true understanding of the history of the earth, and the formation of the fossil record, the things that are found are just as they would be expected. The sifting capacity of water would always tend to deposit sediment in layers that are in complete conformity with each other—just as the rock layers are found, with no hint of time or erosion between them. The fossilized organisms that are found in the rock layers would tend to be found situated according to their ecological environments, as well as their natural buoyancy characteristics (in general with smaller organisms in the bottom layers, and larger animals, including the dinosaurs, in the uppermost layers). Certainly, because of the cataclysmic nature of the flood, it would also be expected that not all organisms or artifacts would be found in perfectly ordered layers. It would furthermore be expected that some land masses would arise, and others would sink. In fact, it would be highly unlikely that there were not evidences of human habitation now to be found under the oceans. Neither geology nor archaeology can be properly accomplished or interpreted without a true knowledge of the Noahic Flood. On the other hand, if there was no global flood, then the so-called anomalies that are found are indeed inexplicable.

Endnotes

1 Lyell, Charles. Cited in Berlitz, Charles. *Mysteries From Forgotten Worlds*. New York: Del Publishing Co., Inc., 1972, 13–14.
2 Cremo, Michael A., and Richard L. Thompson. *Forbidden Archeology: The Hidden History of the Human Race*, rev. ed. Los Angeles: Bhaktivedanta Book Publishing, Inc., 1998, 809.
3 Ibid., 798–806.
4 "Relic of a Bygone Age." *Scientific American*. (June 5, 1852), cited in Cremo and Thomson, 798.
5 Author's note: Tubal-Cain is first mentioned in Genesis 4:22 as someone who "forged all kinds of tools out of bronze and iron." Tubal-Cain was just a few generations removed from Adam and clearly lived during Adam's lifetime. The so-called bronze age and iron age actually took place at the very beginning of creation, and the sedimentary layers do not distinguish separate time periods, but a flood.
6 Cremo and Thompson, 802–03.
7 Ibid., 808.
8 Cremo, Michael A., and Richard L. Thomson. *The Hidden History of the Human Race*. Badger, California: Govardhan Hill Publishing, 1994.
9 Helfinstine, Robert F., & Jerry D. Roth. *Texas Tracks and Artifacts: Do Texas Fossils Indicate Coexistence of Men and Dinosaurs?* 1994.
10 Buagh, Carl E., and Clifford A. Wilson, Ph.D. *Dinosaur*. Orange, California: Promise Publishing Co., 1987, Q.
11 Ibid., 92.

12 Barnes, F. A. "Mine Operation Uncovers Puzling Remains of Ancient Man." *Times-Independent*. (June 3, 1971), cited in Morris, *That Their Words May Be Used Against Them*, 253–54.
13 Ibid., 254.
14 Barnes, F. A. "The Case of the Bones in Stone." *Desert*. (February 1975): 36–39, cited in Morris, 254.
15 Berlitz, Charles. *Doomsday 1999 A. D.* New York: Doubleday & Company, Inc., 1981, 192–93.
16 Ibid., 193.
17 Ibid.
18 Ibid., 124.
19 Ibid., 126.
20 Ibid., 127–28.
21 Ibid., 128
22 Ibid., 134.

Chapter 15

THE PRE-FLOOD PYRAMIDS

Perhaps some of the greatest "archaeological anomalies" of all are the Great Pyramids of Egypt. If you have read this far in the book (and not just skipped around), it should be clear that the past occurrence of a global flood necessarily affects our understanding of virtually every historical or scientific discipline. The disciplines of history, archaeology, geology, paleontology, anthropology, indeed, even such things as biology and astronomy, cannot be properly studied or interpreted apart from a recognition and understanding of the flood. It was the defining event of earth's physical history. Certainly, geology, paleontology, and archaeology cannot be done (as they have been) without understanding and acknowledging the overwhelming consequences of a world inundated by water. With that in mind, I believe that the Great Pyramids of Egypt are not "Egyptian" at all, and can be better explained by understanding biblical history and the veracity of Noah's Flood.

The Great Pyramid is one of the earth's greatest man-made enigmas. Numerous features have baffled investi-

The Great Pyramid of Giza, Egypt and its sister pyramids.

gating archaeologists, Egyptologists, and historians in general. The Great Pyramid ("Khufu," as well as its two sister pyramids, "Menkaure" and "Khafre," and the Sphinx) seems to have required inexplicably advanced architectural, engineering, and construction techniques, as well as mathematical and even astronomical knowledge, that are both unprecedented historically and seemingly anachronistic. Other lesser pyramids, built just a century or two after the Great Pyramid—according to conventional chronology—have long since collapsed. The blocks used to build these lesser pyramids, unlike those of the Great Pyramids, are such that a few men could conceivably handle. In short, the Great Pyramid and its sisters should be the results of thousands of years of accumulated knowledge and experience, and yet they apparently have no precursors and no comparable successors. The knowledge and technology required for such building feats were seemingly lost as rapidly as they appeared. Such apparent anachronisms beg for adequate explanations.

In addition, several features of the Great Pyramid seem to indicate that it was never fully completed or utilized for its intended purpose. No inscriptions, treasures, or mummies were found inside. "Air shafts" neither extended to the outside nor finished their path to the inside. Other more subtle features may imply the unfinished nature of the pyramid as well.[1] Its very immensity seems to preclude the possibility that it was completed during the lifetime of one king or pharaoh. Furthermore, recent studies indicate that it, as well as the nearby Sphinx, may well have been subjected to extensive water damage. The surrounding environment of the Sahara Desert, from the time period assigned to the construction of the Great Pyramid up until the present time, makes this highly unlikely, if not impossible. To date, in my estimation, no single explanation has adequately accounted for all the disparate features and anomalies surrounding the Great Pyramids.

The Pre-flood Pyramids

Several books have been written of the pyramids and the mysteries surrounding them. Only recently, however, have some begun to question the overall chronology of their construction and place in history. In *The Fingerprints of the Gods*, Graham Hancock elucidates numerous inexplicable aspects of the Great Pyramid and conjectures that the pyramid is actually far older than any currently accepted chronology would allow, and evidence for a long-lost "super race."[2] Such a conclusion, however, flies in the face of current evolutionary reasoning. As Egyptologist and author John West points out, "The fact that the technology involved in making [the Sphinx and the Great Pyramid] is in many ways almost beyond our own capacities, *contradicts the belief that civilization and technology have evolved in a straightforward, linear way* [italics added]..."[3]

Other authors have followed similar veins, but most of the more recent hypotheses attempting to rewrite the history of the pyramids are replete with speculations that have little or no historical framework. Even by Hancock's admission, any reconstruction is just guesswork: "The imagination," he states, "is inclined to roam freely over such data in search of an explanation—and all such explanations can only be guesswork."[4] "Guesswork," unless, that is, there can be found a historical framework that can both incorporate and explain the data. I believe that such a framework exists within the biblical account of the Noahic Flood.

The Great Pyramid is an immense structure by any standards. It covers over thirteen acres at its base and is estimated to weigh over six million tons. By any stretch of the imagination, building such a structure would have required enormous effort and almost inconceivable resources—and for what purpose? It is virtually inconceivable that such tremendous effort was undertaken for a building that was to be the sealed tomb of a long-dead and long-forgotten pharaoh. In addition, once covering its surface was a mirrorlike cladding made from an estimated 115,000 highly polished casing stones, each weighing approximately ten tons.[5] Most of this facing was shaken loose by a massive earthquake in AD 1301. Even so, when W. M. Flinders Petrie first examined the remains in modern times, he was stunned to find that these ten-ton facing stones were perfectly fitted together with tolerances of less than one one-hundredth of an inch. As he stated, "Merely to place such stones in exact contact would be careful work, ... but to do so with cement in the joint seems almost impossible; it is to be compared to the finest opticians' work on a scale of acres."[6]

Furthermore, the building seems to be perfectly aligned and positioned, for reasons that continue to baffle modern researchers. The faces of the pyramid are perfectly aligned to north, south, east, and west with maximum divergence of less than .015 percent. Modern structural engineers cannot begin to fathom why such perfection should have been attempted, let alone how it could have been achieved. Hancock asserts that if the "error" had been as much as 1 percent, it is highly unlikely any layman could have visibly detected the difference between such an error and the actual divergence. And yet 1 percent error is almost seventy times the maximum error found in the construction of the pyramid. The corners of the pyramid are perfect right angles, again with much smaller tolerances for error than anything built by modern technology. The base is a perfect square, and the ratio of the height to the perimeter seems to have been calculated to be the same as that between the radius of a circle and its circumference ($1:2\pi$).[7] In order to maintain this ratio, the angles of the sides had to be precisely maintained at the unlikely figure of 52 degrees.[8] Because of such phenomena, recent prognosticators have even postulated that the Great Pyramid may have been a projection map of the Northern Hemisphere of the earth.[9] Whether or not that is the case, there is both historical and persuasive scientific evidence that something was intended by the size and shape of the pyramid, and there is much more to the construction of the pyramid than has been previously conceded.

The Pre-flood Pyramids

Charles Berlitz, author of several admittedly speculative books regarding "unusual" phenomena on earth, nevertheless records what may be corroborating evidence for such a conclusion. According to Berlitz, even though historians accept Pharaoh Cheops (Khufu) of the IV Dynasty as builder of the Great Pyramid, older Coptic traditions disagree. As he records, the Copts were the "purest descendants of the ancient Egyptian stock," and the Coptic tradition maintains that the Great Pyramids existed centuries before Khufu and he may simply have repaired them.[10] Berlitz further maintains that according to "Masoudi, a medieval Coptic historian, the two greatest pyramids ... were built by Surid, one of the Kings of Egypt *before* [italics in the original] the flood, who built them as a result of a prophetic dream wherein 'the sky came down and the stars fell upon the earth.' His interpreter of dreams, when queried, predicted that 'a great flood would come ... King Surid thereupon ordered the two pyramids to be built and to be recorded through their walls all the secret sciences together with knowledge of the stars as well as all they knew of mathematics and geometry, so that they would be a witness for those who would come after them."[11] Such an explanation dovetails nicely with some of the current discoveries being made of the geometry and architecture of the pyramids, and has the added benefit of explaining the purpose of their massive size—they were built precisely to survive a global flood. Recall also that Noah built the ark as a result of a specific prophetic word from God that the world would be destroyed in a flood. Whether it took Noah 120 years or 500 years, it is apparent that many would have known of his undertaking. Word, both of the flood prophecy, and Noah's task, would likely have spread to many parts of the earth.

Immanuel Velikovsky[12] wrote a book over fifty years ago called *The Earth in Upheaval*, which demonstrated a large number of geological and archaeological evidences for cataclysmic, destructive events having taken place on earth. Velikovsky did not recognize or accept the biblical history of one global flood, so he speculated that many, different catastrophes had occurred. Nevertheless, the book was attacked and ridiculed in academic and scientific circles because of its implications for both the potential shortening of the time period required for the formation of many of the earth's geological features, as well as the potential elimination of cherished evolutionary concepts.

Velikovsky also recognized many anomalous features of the Great Pyramid and its environs. For one, the Sahara Desert, which stretches from the Nile to the Atlantic Ocean and covers somewhere in the neighborhood of four million square miles, is in itself difficult to explain. Where did this vast ocean of sand come from if it was not deposited by a retreating sea? Apart from such an inexplicable phenomenon, how would it have been possible in the midst of such an overwhelming stretch of desert to build a structure like the pyramid? Humans and animals need to be watered and fed. Modern estimates have tried to place as many as a hundred thousand men working on the pyramid at any one time, but such an army needs to be housed and fed and, especially, watered. Why begin such an unbelievable construction project in one of the most inhospitable places on earth?

Furthermore, Velikovsky points out, rock carvings and paintings found in the surrounding desert seem to point to a much different environment than currently exists in this area. "There are rock paintings of war chariots drawn by horses 'in an area where these animals could not survive two days without extraordinary precautions.'"[13] In addition, there are drawings that appear to be of extinct animals that not only are not supposed to have existed in the times historically attributed to the Great Pyramid, but in fact could not have existed in desert conditions. "The extinct animals in the drawings suggest that these pictures were made sometime during the Ice Age; but the Egyptian motifs in the very same drawings suggest that they were made in historical times."[14]

The Pre-flood Pyramids

Further anomalies are encountered when one begins to attempt to explain building procedures. It is estimated that somewhere around 2.3 million blocks were used to build the Great Pyramid, with an average weight of several tons each. There are 203 separate levels or layers of rock, but rather than diminishing in size as the height increases, in order to ease the tremendous burden of raising huge masses of rock and perfectly positioning them, the actual structure is far more complex. Over the course of the first eighteen levels, the blocks are seen to diminish in size and weight as might be expected, and they average somewhere between two and six tons apiece. But at the nineteenth course, they suddenly increase in size to an average of ten to fifteen tons apiece. Each of these massive blocks had to have been, not only perfectly aligned and fitted with each of the adjacent, underlying, and overlying blocks, but also raised over a hundred feet in the air just to achieve this level (and eventually to almost five hundred feet near the apex). It may seem perfectly plausible to simply continue adding workers to make such a feat possible, but there comes a point when the sheer number of people required becomes a problem greater than any potential benefit. If you can visualize using people to lift a compact car, you will have some idea of the difficulties encountered in manipulating even one block into place. Suppose, for instance, you use ten people to try and lift a three-thousand-pound car (see if they could carry it up the side of the pyramid at an angle of 52 degrees to put it in place!). Each must lift an average of three hundred pounds. That probably won't work for long, so increase the number of people to one hundred, so that each person has to lift only thirty pounds. Terrific, but where do they stand and what do they hold on to? And then realize that a fifteen-ton block (which is by no means the largest of the pyramid blocks so lifted) is the equivalent of ten three-thousand-pound cars stacked on top of one another, and you begin to understand the tremendous difficulties any explanation faces. Because now, even if you somehow fit one hundred people around one car, each has to, once again, lift three hundred pounds and carry that weight up the side of the pyramid. And, as if that were not enough, the thirty-fifth course consists of blocks much more massive than those preceding it.[15] Adding greater numbers of people may sound sufficient, but in reality it does little to explain the engineering feats required to lift the blocks into place or even to so precisely carve them in the first place. The truth is that no current explanation of how such a feat could have been achieved is remotely adequate. Those historians who have speculated about "scaffolding" and "ramps of sand" are simply ignorant of the complete inadequacy of such theories from an engineering and construction standpoint.

Most current explanations have the Egyptians building ramps of dirt that were later removed. In reality, such an explanation falls woefully short for numerous reasons. As Hancock states, "To carry an inclined plane to the top of the Great Pyramid at a gradient of 1:10 would have required a ramp 4800 feet long and more than three times as massive as the Great Pyramid itself (with an estimated volume of 8 million cubic meters as against the pyramid's 2.6 million cubic meters). Heavy weights could not have been dragged up any gradient steeper than this by any [known] normal means. If a lesser gradient [more gradual incline] had been chosen, the ramp would have had to be even more absurdly and disproportionately massive."[16]

In reality, any such ramp made of lesser materials than the pyramid itself would collapse under its own weight. Therefore, to build the pyramid by use of ramps and manpower would actually require the building of the equivalent of three additional adjacent and complementary pyramids all of which needed to be removed and disposed of afterward.[17] This can hardly be an adequate explanation when one of the greatest dilemmas in explaining even one such structure is how the construction could possibly have been accomplished during the life span of one person. And if it was not built by, or for, one person, then it becomes even more difficult to explain how such a project could have been maintained for centuries after the initiator had passed from the scene.

The Pre-flood Pyramids

Perhaps even more enigmatic than the apparent perfection of angles and corners and ratios and relationships and positioning, is what was found inside the pyramid. According to historical records, Caliph Al-Ma'mun, the Muslim governor of Cairo in the ninth century AD, hired a team of workers to tunnel their way into the pyramid to see what they could find. No entrance was visible or known, so the only option was to quarry their way into the side. Amazingly, their tunnel happened to join up with one of the internal passageways that descended from the original concealed doorway in the northern face. The vibrations from their excavations also dislodged a block of limestone from the ceiling of the corridor. Inside this niche, they discovered the beginning of an ascending corridor. Access was denied, however, by a series of huge granite plugs that were fitted precisely into the corridor and were both immovable and impenetrable. The quarriers had to tunnel around these blocks and it became apparent that, when they eventually reconnected with the ascending corridor, they were about to enter an area never before breached.[18] And yet, once inside the so-called King's Chamber, they found absolutely nothing. No hint of any treasure, no mummy, no hieroglyphics or carvings or paintings or statues; nothing but an empty granite box later said to be a sarcophagus. As Hancock states, "No other proven burial place of any Egyptian monarch had ever been found undecorated."[19] No hint of an earlier break-in was ever discovered, either in the pyramid itself or in historical records. Though another shaft of nearly 160 vertical feet was eventually discovered in the nineteenth century, it was not only well hidden, but sealed with an impenetrable mass of debris. Whether or not this shaft had been open and accessible to earlier centuries of grave robbers remains unknown, but it appears highly unlikely that large amounts of treasure and/or pharaohs' mummies could ever have been removed by that route. The evidence seems to indicate that these chambers were empty from the very beginning.[20] One explanation that seems to have been conspicuously avoided is that catastrophe overtook the builders and the Great Pyramid was never completed, nor utilized for its intended purpose. That is, unless its primary intended purposes were enigmatically hidden in the very size, dimensions, and shape of the structure itself.

Two other features of the Great Pyramid that also cry out for explanation are the so-called air shafts in the "King's" and "Queen's" Chambers and the water damage or erosion to the exterior. In the King's Chamber were discovered two long, narrow shafts that extended to the exterior of the pyramid. Egyptologists assume that these were somehow intended as ventilation shafts (all mummies need to breathe, after all), and yet, in the secondary Queen's Chamber, no such shafts were originally found. When it was assumed that they were there, but hidden, a search was initiated that eventually disclosed two such shafts. Both of them stopped several inches short of making egress into the chamber, however, and neither of them extended to the exterior of the pyramid. These rectangular shafts are only circa eight by nine inches—too small for use as human access.[21] So, why would the builders of these shafts leave the last five inches to the Queen's Chamber unfinished? What purpose could they possibly have had in the first place, since they did not even access the outside of the pyramid? Is it possible that whatever their purpose, they simply were not completed when work was halted by the cataclysmic destruction of the civilization that was building them?

In recent years, geologists have placed even more enigmas on the table regarding the Great Pyramid and the Sphinx. Robert Schoch showed in 1992 that the "body of the Sphinx and the walls of the Sphinx ditch are deeply weathered and eroded ... This erosion is a couple metres thick in places, at least on the walls. It's very deep, it's very old in my opinion, and it gives a rolling and undulating profile ..."[22] He concluded that it was "clearly rain precipitation that produced these erosional features."[23] The erosional features, he noted, could only have been formed by enormous quantities of rain and water, something that is obviously quite foreign to the current Giza

The Pre-flood Pyramids

plateau. His conclusion, and that of West and Hancock as well, is that the Great Pyramids and the Sphinx are the products of a civilization centuries—perhaps millennia—older than any recognized reconstruction of Egyptian history. He believes that these great monuments show clear, scientific evidence of having been extensively weathered by water that he attributes to a much earlier "pre-Ice Age" epoch.[24] As West states, "All that I know for sure on the basis of our work on the Sphinx is that a very, very high, sophisticated civilization capable of undertaking construction projects on a grand scale was present in Egypt in the very distant past. *Then there was a lot of rain.* [italics added] Then, thousands of years later, in the same place, pharaonic civilization popped up already fully formed, apparently out of nowhere, with all its knowledge complete. That much we can be certain of."[25]

R. A. Scwhaller, a French mathematician who theorized about the tremendous mathematical capabilities of these pyramid builders, also concluded that "a great civilization must have preceded the vast movements of water that passed over Egypt, which leads us to assume that the Sphinx already existed, sculptured in the rock ... whose leonine body ... *shows indisputable signs of water erosion* [italics added]."[26] Again, how is it possible that these massive structures were built—and yet perhaps left unfinished?—with such inexplicable and anachronistic technology in the midst of a desert, and yet eroded by vast amounts of water (that no longer exists!) by a highly advanced civilization that seems to have disappeared with no other trace?

As I stated previously, I believe that every enigma presented by the Great Pyramid and the Sphinx is best understood in light of the Noahic Flood of Scripture. As Schwaller said, "A great civilization must have preceded the vast amounts of water that passed over Egypt."[27] As Velikovsky also instinctively realized, "the conflict between the historical and the paleontological evidence, and of both of them with the geological evidence, is resolved *if one or more catastrophes intervened.*[italics added]"²⁸ Even Hancock is forced to speculate that the incredible civilization that was capable of building such structures as the Great Pyramid and the Sphinx, must have come to an end, "extinguished by some sort of massive catastrophe."²⁹ If the Great Pyramid is actually a structure that was built before the flood, virtually all of its most inexplicable features fall neatly into place. There was a highly intelligent and very sophisticated civilization in place for 1,656 years before the flood (according to the Hebrew biblical chronology). Most information from, and evidence of, this civilization was destroyed in the flood (the complete destruction of this corrupt and immoral civilization was, after all, the very reason for the flood in the first place). The technology required to build the Great Pyramid, for the most part, did not survive the flood itself, even though the pyramid did.³⁰ Noah was undoubtedly highly intelligent and capable himself, but he did not carry with him all the previously existing technologies (the same would be true of any one individual of today, no matter how intelligent). His job was not to build a pyramid, but to build a ship. Even today, it has been claimed by modern ship builders, a wooden ship could not be built with structural integrity the size of Noah's Ark (maybe not, by today's technology ... but Noah did it!).

Another feature of the Great Pyramid, which dovetails well with the idea of a pre-flood structure, is the length of time that seems to be required for its construction. Many difficulties have arisen with attempts to explain the construction of the pyramid during the reign of any one pharaoh. Suppose, just for the sake of argument, that there is no problem in lifting fifteen-ton blocks hundreds of feet into the air and perfectly fitting them to every other block in the structure. Forget about all of the design and technology difficulties to overcome.³¹ Forget about transporting such tremendous blocks through the desert, and providing sustenance and housing for untold thousands of workers. If we tackle only the problem of placing the blocks in position, the immensity of the dilemma is still

The Pre-flood Pyramids

overwhelming. Suppose one block could be properly raised and fitted every hour on the hour, seven days per week.[32] If we assume a ten-hour workday, it would still require over 650 years to build the Great Pyramid alone. And that is completely neglecting all the incredible hidden aspects of the pyramid: interior rooms, tunnels, secret chambers, "ventilation" shafts, and so forth, all would require far greater periods of time to both design and construct. To illustrate by only one example, the so-called King's Chamber in the Great Pyramid consists of walls made with one hundred blocks weighing seventy tons apiece, and a ceiling made of nine blocks weighing fifty tons apiece, all in a chamber 150 feet above the ground![33] Even if you allowed the Egyptian workers to somehow work throughout the night (did they also have stadium lighting?) in order to work twenty-four hours a day, it still is a monumental task that would appear to require several hundred years at least. Now, however, recall that according to the biblical pre-flood genealogies in Genesis, most antediluvian men lived at least nine hundred years! Projects that would be impossible to accomplish within the life span of one man (who might be pharaoh for at most fifty years) become entirely possible if a period of five hundred years or more is available. Someone who anticipated living for over nine hundred years could very well undertake a project that might last for several hundred years. For instance, it is very possible that Noah had several centuries in which to complete the Ark.[34] Recall also that Scripture speaks of the Nephilim, the "heroes" of old—undoubtedly giants of intellect as well as stature. Just imagine how much knowledge one person could accumulate with a life span of nine hundred years! Such a possibility makes the technological achievements of the pyramids entirely feasible. Placing the Great Pyramids and the Sphinx as pre-flood creations also explains what happened to such knowledge and technology—they were largely destroyed just as the Bible indicates. The flood also helps to explain the desert, as well as the placement of the pyramid. It was not built in the middle of a desert; the desert is the aftermath of the flood that destroyed the people and the animals that once lived there. This has the further benefit of nicely explaining the artwork of now-extinct animals—these animals were around and plentiful before the flood.

Unlike the so-called Ice Age (see chapter "The Myth of the Ice Age"), numerous flood accounts exist throughout the world in diverse and apparently unrelated cultures, and this includes the Egyptians. Egyptian legend, according to Berlitz, records a massive, cataclysmic flood sent by Ra, the sun god, at the "end of the rule of the God Kings who reigned before the Flood [italics added] ..."[35] Furthermore, states Berlitz, "the Great Pyramid of Gizeh was said to have been constructed shortly before ... the flood, not as a tomb but *as a permanent and indestructible storehouse of knowledge that was literally built into its dimensions and orientation* [italics added]."[36] Further ancient sources refer to the Great Pyramids as structures made by the great kings who lived before the flood.[37]

The global flood of Scripture is an event that is either ignored as fiction, or "mythologized" as irrelevant in virtually every modern attempt at reconstructing the past. And yet its occurrence would necessarily mean it was the primary agent of not only the geological and fossiliferous features of the earth, but the destruction of civilizations as well. Any reconstruction of the past that denies or ignores the agency of the flood would have to be relegated to the fiction aisles of bookstores and libraries. Ultimately, even the harshest critics of the biblical account must recognize that if the Bible is true history, then only the understanding and acceptance of the biblical framework enable data to be properly interpreted.[38] Lest anyone forget, Scripture warns that in the last days scoffers will come and they will deliberately forget that a flood once inundated the entire world! (II Peter 3:3–6.)

Endnotes

1 Hancock, Graham. *The Fingerprints of the Gods*. New York: Three Rivers Press, 1995, 323. As Hancock also notes, for instance, of the Queen's Chamber, that although the walls and ceilings were elaborate and even elegant, the floor "was the opposite of elegant and *looked unfinished* [italics added]." 323.

The Pre-flood Pyramids

2 Ibid., 4.
3 Ibid., 426.
4 Ibid., 413.
5 Ibid., 290.
6 Petrie, W. M. Flinders. *Traveller's Key to Ancient Egypt*. 90, cited in Hancock, 290.
7 The height has been calculated as 481.3949 ft. and the perimeter as 3023.16 ft. Cf. Hancock, 319. However, since the original surface has been lost, some doubt must remain in assigning "perfect" values.
8 Ibid., 276–280. It seems evident that this angle was specifically chosen to accomplish certain aspects of the pyramid, since other choices, like 45°, would clearly be simpler to calculate as well as to maintain from a construction standpoint. Other researchers, however, have seen significance in possible angles of 51°50.6', 51°49.6', and 51°51.2' and come to alternate conclusions of the actual ratio and the reason for the slope of the sides—such as the "golden section". Cf. Ley, Willy. *Another Look at Atlantis*. New York: Galaxy Publishing Corporation, 1969, 39–45. What does seem clear is that the angle was intended by the builders for a specific purpose and not randomly achieved.
9 Ibid., 434–36.
10 Berlitz, Charles. *Doomsday 1999, A.D.* New York: Doubleday & Co. Inc., 1981, 19.
11 Ibid., 19–20.
12 Hancock seems to have followed many of the same sources as Velikovsky, and in some cases, to have reached similar conclusions; but with little or no recognition of Velikovsky. Other than an obscure footnote, he makes no mention of Velikovsky in the text. Whether this is coincidental or intentional is unclear. Velikovsky was much maligned in his time, and perhaps Hancock hopes to avoid a similar fate.
13 Velikovsky, Immanuel. *The Earth in Upheaval*. Garden City, New York: Doubleday & Company, Inc., 1955, 94.
14 Ibid. (Author's note: by implication, then, the "Ice Age" and historical times must have been one and the same—the "Ice Age" as popularly taught did not occur; a global flood did.)
15 Hancock, 282–83.
16 Ibid., 285.
17 Ibid.
18 Ibid., 296.
19 Ibid., 301.
20 Ibid., 299–300
21 Author's note: Interestingly enough, in 1993, when a German robotics engineer was hired to clear out these shafts and even install fans to improve ventilation, his remote-controlled robot, 200 feet up this tunnel from the Queen's Chamber, discovered a previously

unknown and inaccessible sealed chamber closed with a limestone door and metal fittings. Cf. Hancock, 320.
22 Schoch, Robert. Debate: "How Old is the Sphinx?" AAAS, Annual Meeting, 1992, cited in Hancock, 421.
23 Hancock, 421.
24 Ibid., 357.
25 West, John. Interview with Hancock cited in Hancock, 427.
26 Schwaller de Lubicz, R. A. *Sacred Science: The King of Pharaonic Theocracy.* Rochester, Vermont: Inner Traditions International, 1988, 96, cited in Hancock, 419.
27 Ibid., 419.
28 Velikovsky, 94.
29 Hancock, 427.
30 Author's note: When I lived in Utah from 1983 to 1985, I witnessed firsthand the results of a flood that inundated an entire town. A landslide blocked the path of a river that then flooded the entire valley, creating a lake where there had once been a small town. Many months later, when engineers eventually drained the lake, it was amazing to see some wooden houses still standing structurally intact. Others that had been in the path of onrushing water were completely demolished and even buried in layers of sediment. In the Noahic Flood, obviously vast layers of sediment were deposited—especially in the regions surrounding the opening of the "fountains of the great deep." Yet, it is to be expected that in some areas the waters would have risen less destructively (to structures), and a building with the massive structural integrity of the Great Pyramid could have remained intact. Obviously, if the Great Pyramid were built upon layers of sedimentary rock replete with flood-deposited fossils, this could falsify my interpretation, but to my knowledge, such is not the case.
31 According to Hancock, there are only two construction cranes in existence in the modern world that are capable of manipulating some of the pyramid blocks into place; and it would take approximately six weeks to set up and move one block! Cf. 342.
32 Author's note: Anyone familiar with construction efforts involving such enormous objects—like bridge spans or even iron beams for high-rise buildings—can quickly appreciate the overly conservative nature of such an estimate.
33 Ibid., caption for photo #44.
34 Though many commentators state that Noah built the ark in 120 years, based on Genesis 6:3, this is not necessarily clear from the text itself. An alternate rendering of the same verse indicates that it is the life span of an individual man that is going to be limited by God to 120 years, and not the amount of time left before the flood. That this is a possible meaning of the verse is attested to in the NIV footnotes and bolstered by the fact that current human life spans do indeed appear limited to a maximum of 120 years. [For a

The Pre-flood Pyramids

more detailed discussion of this subject, refer to chapter 12, "The Scriptural Advent of Animal Carnivorousness," pp. 231–234, and endnotes 7–11 from that section.]
35 Berlitz, 162.
36 Ibid., 169.
37 Ibid., 193.
38 Since, for instance, as the biblical genealogies make clear, mankind has existed for only around six thousand years, any speculation about "pre-historical" man existing fifty thousand (or more) years ago is simply untenable.

Chapter 16

THE 2003 SEARCH FOR NOAH'S ARK

This year, I flew from LAX on a Sunday evening to London and then on to Istanbul. It has always been a toss-up whether it is best to leave in the evening and arrive in Istanbul even later the following evening—missing the day in between—or to fly in the morning, arriving the following morning having missed the evening in between (Istanbul is ten hours ahead of L.A. in time zones). The advantage of leaving at night is that you've had a long day already and your chances of sleeping on the plane are better. If, however, you do manage to catch some sleep on the plane, you arrive in Istanbul late at night and you are not tired enough to sleep again—which is exactly what happened to me this year. On the other hand, the advantage of leaving in the morning and arriving the next morning, not having slept in between, is that you must keep going all day, and by evening you are undoubtedly exhausted enough to get some sleep. Of course, that is also the disadvantage of doing the same—you are exhausted, and must go for twenty-four hours or more without sleep.

In any case, I arrived in Istanbul at roughly 11:00 p.m. local time (9:00 a.m. by my body clock) and had to pay for my Visa (now $100!), change $300 into Turkish lira (roughly four hundred million lira this year!), get my luggage, call the Hotel Airport Inn (which Richard Bright had recommended to me), get picked up, and check into the hotel all before midnight—but, as tired as I should have been, it was morning for me and I could not sleep. I was, however, very pleased to learn that the room had its own air-conditioning unit—a rarity in Turkey (at least in the establishments that I frequent). That alone made the stay worth the $60 fee.

The 2003 Search For Noah's Ark

In the morning, I had to catch a cab to the airport still wondering if I would have a reservation on a plane to eastern Turkey. The reservation from Istanbul to Van, made in America, had been quoted to me as $560 by the travel agent. That seemed exorbitant to me as I usually find travel within Turkey to be quite reasonable (especially by bus). When the agent told me that I could make my reservations and then pay when I got to Turkey, I decided to risk it. When I arrived at the very sparsely populated airport at 8:30 a.m., there was one ticket window open and no one waiting. It proved to be no problem to confirm my reservation and the ticket cost me only $225 at the Turkish end of things—less than half the price from America.

From Istanbul, I flew to Van, arriving at roughly 2:00 p.m., not entirely sure how to catch a bus from there to Dogubayazit. Thankfully, Richard Bright had called Mehmet, a friend of his in Van, and asked him to meet me at the airport and give me a ride. Somehow, without a picture to go by, Mehmet was able to pick me out of the crowd of swarthy Kurds with no problem. After being picked up, I breathed a large inward sigh of relief in believing that I would now have a comfortable two-hour car ride to Dogubayazit, rather than tramping around Van in search of a crowded mini-bus for a three-hour ride to same. Unfortunately, I did not realize that Mehmet was from Van, and not from Dogubayazit.

As it turned out, my disappointment was palpable when Mehmet drove me to the minibus station and bought me a ticket to Dogubayazit. My one consolation was that he told me that he would speak to the driver and that once they let off the other passengers in Dogubayazit, the driver would take me the additional five miles up the hill to Murat Camping where Richard was staying. That would save me the hassle of lugging my backpack and suitcase around town looking for a cab to take me up the hill.

Of all the ways to travel in Turkey, by far the least commendable method is the "minibus." Though cheap (from Van to Dogubayazit, roughly one hundred miles, the cost

Dinosaurs on the Ark

was less than $6), it can be challenging to say the least. Unlike the buses, which are spacious and air-conditioned, the minibuses are usually cramped, crowded, hot, and quite dirty. In this particular van we had up to eighteen passengers (not counting the driver) crammed into a space designed for, perhaps, twelve. Kurdish hygiene may be fine under normal conditions, but crammed into such a small and warm place, the aromas endured can bring a tear to your eye. The only "fresh" air available is when you are moving, and on a hot day, it makes little difference. All manner of foodstuffs, and even animals, are piled on top of and into the van in every conceivable corner.

On this particular trip, we started out with seventeen passengers, made our way through the crowded havoc of the Van city streets, drove about five miles out of town, and pulled over to let off two passengers. I breathed an initial sigh of relief when I could have a little more breathing room. But when the van then turned around and headed back into town to pick up two more replacement passengers, I almost cried. How many times after that we stopped and either let off passengers or picked up more is somewhat of a blur, but at the end, we had more passengers than we started with.

At one point in the trip, I happened to hear one of the passengers speaking in English (a rarity in that neck of the woods), and amazingly enough, he was speaking about evolution. In fact, he was speaking to the one other passenger on the van who was clearly a "tourist" like myself (Japanese). I tuned in just in time to hear him say that, just the day before, his friend told him that he did not believe in evolution. He seemed incredulous that in this day and age, someone might not believe in evolution. He stated simply that evolution was a proven scientific fact! My ears were burning, but there seemed no way for me to intrude on a conversation in a crammed minibus with someone two rows back. I decided that if this was indeed a divine appointment, the Lord would somehow have to put us together in a different place and time.

Amazingly enough, ten or fifteen minutes later, the van stopped to pick up more passengers ("Are they going to sit on my lap?" I wondered.). The driver got out of the van and came around to the side door to survey the hopelessly crowded situation and motioned for me to move into the back row of the van—right between the two men who had been having the "discussion." I prayed about how to proceed, and then started by asking the Turkish man how he had learned such good English. He proceeded to tell me that he had just graduated from college and he was on his way to Dogubayazit to finish out his military requirement. He had also had a job that required him to use English. I mentioned that I overheard him say that "evolution was a fact" and I wondered what led him to believe that. He smiled and eagerly proceeded to tell me that science had showed it to be true. "That's funny," I countered, "in America I study and teach about evolution, and I have come to the opposite conclusion—namely that science shows evolution to be impossible." He seemed taken aback, but also a bit curious, so we began to discuss it. I asked him to tell me what "proof" scientists had for evolution and he replied that, for instance, fossils showed that apes had evolved into men. I then gave him several examples of such "proofs" that were either fraudulent or mistaken. Piltdown man, for instance, was for over forty years said to represent "the best evidence" for human evolution and yet it was eventually revealed to have been falsely manufactured from a few human skull fragments and the jawbone of an orangutan (see chapter "The Fossil Record"). I said that the fossil evidence of human evolution actually only demonstrated that some men desperately wanted to believe in evolution because they did not want to be accountable to God.

I then asked him to imagine that he found a stone statue carved into the likeness of a person. "Would you reason that it had been formed by natural processes of erosion?" I asked. He admitted that he would know that it had been formed by a person of intelligence. "Which is more complex and intelligent," I asked, "the statue, or the person who carved it?" The person of course, he intimated. "So," I said, "how is it possible that you can look at a simple stone statue and know unequivocally that it was formed by an intelligent being, but you cannot look at a person—infinitely more complex and intelligent than the stone he carved—and reason the same thing?" Human beings must have been created by a being infinitely more intelligent than ourselves. As brilliant as human scientists may be, they are totally unable to create even a single-celled organism in a lab.

Ahmed (his name) then admitted that he did think there must be a Creator of some sort behind the whole process, but that we were the product of a long process of evolution that this being had put into place. However, I pointed out, what we actually see in nature is not a process of improvement of species, but one of decay and deterioration. I used the example of experiments done with fruit flies—one hundred years'-worth of experiments trying to force fruit flies to evolve have produced many mutational defects—fruit flies with no wings, or no legs, or two heads, etc.—but we still have only fruit flies to show for our efforts. Most of the defective fruit flies are "weeded out" in nature, but that only maintains the fruit fly, it does not improve it or change it in any appreciable way. In fact, the only scientifically observed direction in nature is one of extinction and deterioration, but not of evolution.

He seemed to contemplate this, and then, almost as if he had a revelation, he began to recall to me the ways that they were all brainwashed in the schools of Turkey. He said that they were taught evolution in their biology classes, but they were not allowed to question any aspect of it. Eventually, they came to just accept it. Then he told me that their classrooms were regularly visited by military officers who came specifically to reinforce this concept.

One time, he remembered, an officer had come to his primary school and told each member of his class to pray and ask God for something specific that they wanted right then and there. When nothing appeared in front of them,

The 2003 Search For Noah's Ark

the officer claimed this proved that God did not exist and did not answer prayers.

Another time an officer handed a book to one of his classmates and asked the child to tell him how he knew it was real. "Because I can see it and feel it," the child had answered. "Yes," the officer replied, "and where is God? You cannot see or feel God, because He does not exist. It is only in your imagination." With that, I interceded, "I wish I had been there, I would have asked the officer to show me his brain. If I could not immediately see or feel it, I would be forced to conclude that he did not have one." Ahmed suddenly became animated as he recalled that a girl in his class had actually asked almost that very question. She had said, "Can you show me your intelligence? If not, then you must not have any." In the course of our conversation, I was amazed to learn firsthand what a direct and active role the military has taken in Turkish schools to eradicate belief in God. Later, I was able to confirm this through my friend Salih, as well as from Saim. I felt that my meeting with Ahmed was a divine appointment that confirmed my belief that the military of Turkey has purposely determined to thwart any discovery of the ark.

Ahmed, my new friend, seemed to have come to a new revelation of the intentional nature of his brainwashing, and he was not happy about it. He said, with genuine emotion, "I have never heard these kinds of things before. I am very glad that you talked to me." We then exchanged e-mail addresses so that he could write to me and learn more. Not long thereafter, the minibus pulled in to Dogubayazit and began disgorging its contents—passengers, animals, various foodstuffs, and luggage.

As everyone was unloading, I asked the driver if he was going to continue up the hill to the campground. He appeared to discuss it with someone "in charge" who said, "Yes, yes." Halfway up the hill, however, the driver began making sly references to money. He pulled a wad of bills from his pocket and seemed to indicate that he now wanted something for his troubles. I thought he indicated "two million" lira, which would have been somewhat less than $1.50 in American money. I held up two fingers and said, "Two?" He grinned and nodded and continued up the hill. I suppose if I had refused he would have stopped and let me out where I was. I realized once again that in Turkey things such as cab fare must be settled in advance.

When we arrived at the campground, I handed him a five-million note expecting him to proffer change. He acted disgusted and pulled out a twenty-million note to show me how much he really wanted. It had cost eight million (less than $6) for the more than one hundred mile-ride from Van, and he wanted twenty million for the five-mile ride from town. I refused, put the five million back in my pocket, and walked away. For a time, he followed me around insisting that I pay him twenty million. When I absolutely refused, he finally disappeared, though Saim told me later that he had paid the driver fifteen million. It seems that many Kurds and Turks of eastern Turkey accept dishonesty as both a way of life and part of doing business.

When I first arrived at the campground, I immediately encountered Richard Bright, who was just on his way to meet someone for dinner. He welcomed me with a hug and a grin. It was obvious that he was relieved that I was finally there and things could get moving. He had been in Turkey for over a week, and he was champing at the bit to start. Richard is an airline pilot for Continental Airlines who has, for the last couple of years, made his home-base in Guam. He has been variously maligned and revered in ark research circles, but both opposite sides of the spectrum acknowledge him as the "bulldog" of all researchers. This year was his twenty-third trip to Turkey and somewhere around his fifteenth climb of Ararat (not even he can remember for certain). In the mid-eighties he climbed as part of Jim Irwin's groups with legal permission from the government, and he even managed one of the few authorized photographing flights over the Ahora Gorge in a rented Cessna in 1986. Never mind that they landed at Erzurum Airport an hour too late and were quickly placed under house arrest by the military and all their film and

The 2003 Search For Noah's Ark

photographic equipment confiscated. With Jim Irwin interceding on their behalf, they eventually were released and some of their now-developed pictures were even returned. However, they never received their negatives, and Richard believes that they received far fewer pictures than they took.

I have climbed Mt. Ararat with Richard three times (including this year) and "arranged" several more climbs that never materialized for one reason or another. Richard would be the first to admit that he has made his share of mistakes, and fallen prey to his share of scams in the search for the ark. His chief flaw, if it is one, is that he is too trusting—and too determined to find the ark. But his perseverance and wholehearted devotion have gotten him all over the mountain at a time when very few other researchers have even made it as far as Turkey, let alone to Dogubayazit. He has climbed more times, spent more money, experienced more heartbreak than any other five researchers combined. And every single year that I have been with him, he has sworn unequivocally that he will never come again! Though, in retrospect, I must confess that I have made the same claim on more than one occasion myself.

In fact, I would have to list the four most physically painfully memorable days in my life—in no particular order—as: (1) coming down from Ararat by myself, after leaving Paul Thomson there, in 1999, (2) coming down from Ararat with Richard in 2000, (3) coming down from Ararat with Richard and Paul in 2001, and (4) coming down from Ararat with Richard in 2003 (2002 is omitted simply because I did not make it to Turkey that year). Of those climbs, 2001 stands out as particularly difficult and painful because we had to make the trek down from nearly seventeen thousand feet—after exploring and photographing above the Ahora Gorge for half a day—all the way back to Dogubayazit in the same day, and much of the descent was in pitch-blackness well into the wee hours of the morning. But this year would also prove to add its own unique flavor to the memories.

The first evening that I was at camp, I happened to run into Mustafa Ozturk, the very man who told me four years ago that he had seen Noah's Ark "once in his life." Meeting him was a surprise as I had heard two years ago that he was in Istanbul running from the law. Saim later told me that Mustafa had returned to serve his prison time and had served only nine months of an eleven-month sentence—something apparently unheard of in Turkey. In any case, Mustafa summarily invited Richard and me to join him and his friends for supper around their campfire that night. Richard, at first, was loath to join them because he was working hard to keep his sojourn in eastern Turkey a secret. However, since it was obvious that our presence was already well known, we decided that it would be the better part of valor to join them. We also thought that there might be an opportunity for me to discuss with Mustafa his earlier promise to one day take me to the ark "for no money, because you are good people." Consequently, we later walked up the hill in the dark and found their campfire where four men were gathered around. Saim, Ashygul (a girl who helps Saim with his tourism business on an ad hoc basis), Richard, and I were quickly made into the guests of honor. In front of us, they placed a large pan that had been simmering in the fire. It consisted of chunks of meat and bone, tomatoes, and various other unknown vegetables. Into the mix they then dumped a bucket of yogurt and someone stirred the entire concoction with his fingers. Cleanliness is rarely an issue in eastern Turkey, and all those gathered around immediately began plunging their hands into the mix to look for the pieces they wanted. They would eat over the pan so that if pieces fell back in they could simply be stirred back into the mix and recycled. Bones would be tossed to the side and then greasy, dirty hands would go back into the pot to stir and look for another delectable morsel. I must confess to a more queasy stomach than Richard, and I avoided most participation by explaining that I had already eaten (which was true). Occasionally, someone would hand me, as guest, a special piece they had retrieved from the mixture, and I would be forced to participate or risk totally offending them. Usually, I just ate very slowly (and chewed very well!) on a piece of bread, pretending on occasion to dip it into the mix.

After dinner, they began liberally imbibing the various bottles of alcohol they had also brought along. I believe that they took my refusal to participate as either an insult or a source of amusement (or both, if that's possible). In any case, around 11:00 p.m., I declared my need for sleep and headed back down the hill.

Mustafa insisted on accompanying me, so I decided that it would be a good opportunity for me to renew our discussion about his sighting of the ark. I got the ball rolling by telling him that I did not believe his story. He seemed taken aback and asked me why.

"Because," I said, "everyone in town wants money. Many people tell me they have seen 'something,' or that they will help me find something, but they always want only money. I am looking for an honest person who says, 'I will take you there and if we find nothing you will pay nothing.' I do not have money to give, and I do not want money. If there is a reward for finding the ark, you may keep it all. I only want to find the ark. And if you really knew where it was, you could show me even one picture, or you could tell me how to get there, but you do not."

Mustafa then waxed serious as he explained that he still had my phone number and that "one day" he would call me, but for now, "politics" were too dangerous. The Turkish military, he explained, is very upset with America and George Bush (Sr.) and the potential for American involvement in helping to establish a new "Kurdistan." The military does not want anyone helping Americans.

Our ongoing discussion after that was to no apparent avail. I kept trying to insist that if the ark was shown to the world, the worldwide attention brought to this area could only help the Kurdish situation. Their plight would begin to receive both the attention and protection of the world. I am not sure if I followed all of Mustafa's ministrations

Dinosaurs on the Ark

in his broken English, but what I gleaned from him and from reading between the lines was that he had somehow gained the favor of the government (maybe as an informant for the police or military?). He had a visa to travel to many places outside of Turkey (something that Saim had insisted was next to impossible for a Kurd in eastern Turkey) in order to do business. He now told me that he was wealthy and did not need any money; he even insisted that he would help me if I needed money. In fact, he said, he had plans to build a "mountain house" at thirty-two hundred meters on Mt. Ararat. It sounded like he intended to open some kind of resort there and one day lead tours to the ark. He even wanted to give me a piece of his land to build my own house and join him!

My guess is that whether he knows anything of the real location of the ark or not, he does not currently want to do anything that might jeopardize his new, good-standing with the government. He told me that he could get me a permit from Ankara to climb the mountain (for only $100), but again, it would only be a tourist permit for the south side and not a search permit. I told him that I had already climbed the south side and there was nothing there! I came to find the ark, not to climb the mountain for amusement. Nevertheless, he insisted that he still had my number and would, one day, call me. He was "working on a plan." Me too.

The first full day after my arrival was spent recovering from the trip and going over pending plans and options with Richard. It began to appear that—surprise, surprise!—things were not as in-place and ready-to-go as we had hoped. The following day, we took a car with Saim and drove around the mountain to get a look at the other side. Nothing of any interest was visible in the area of the so-called Western Anomaly (an area of rock-formation that sometimes protrudes from the snow and has been touted by some as Noah's Ark). As we began to see into the Ahora Gorge, the first thing that became apparent was that there was clearly much more snow than two years previously when Richard, Paul, and I had climbed. That seemed to

The 2003 Search For Noah's Ark

dash any hopes that we might find the structure in a previously searched area simply because of reduced snow coverage. If the ark has been able to remain hidden primarily because of snow-coverage and ice conditions, 2003 would not be the year to find it.

Richard, Saim, and I discussed the various options and it again seemed to me that our plans were quite "fluid." In Turkish-speak that often means delays of many days if not weeks. Returning to our room at the campground forced me to begin praying more fervently that the delay would not be weeks. The room that Richard and I shared had an "electric problem," which meant that there were no lights, and since the water heater was electric, no hot water either. For me, staying in eastern Turkey under the best of conditions is not idyllic, but with no showers and no lights—and very little to do as we waited—self-questioning my own mental health increased. The room that we were in did have its own bathroom but that consisted of only a toilet (though no toilet paper—we made do with napkins we borrowed from the restaurant at the campground) and a shower hose attached to the wall. There is no sink or mirror, and the overall aroma and effect can be decidedly unpleasant. However, we each had a bed (a new feature) and clean, seemingly new, sheets and bedspreads—probably bought for the occasion of our arrival. The noises of the campground—mixed with the adrenalin of excitement and nervousness—made sleep hard to come by, even though I came fully equipped with earplugs. They were of no avail, and after a couple of days of sleep deprivation and trying to adjust to the new time zone halfway around the world, I was anxious to do something. The best I could come up with was a hike to town.

The campground where we were staying is about five miles outside of town and a fair increase in elevation. I got up early the following day and risked an ice-cold shower before heading down the hill for Dogubayazit. I was hoping to arrive at the PTT (post office) by 8:00 a.m. so that I could get in a phone call to the States before the waiting line got too long (there is only one public telephone in town). The phone at the PTT is a much cheaper way to call the States than using a cell phone, but it is also a lot more inconvenient. I had borrowed a Turkish cell phone from friend and premier ark researcher John McIntosh before leaving for Turkey, but it needed a new "chip" in order to be reactivated and assigned a new number. Along the way to the PTT, I stopped at an open cell phone establishment to accomplish just that, but it turned out to be more complicated than I had hoped, so I left them with my passport and decided to return after breakfast. I managed to telephone parts of my family in the States, have a cheese omelette at my favorite in-town restaurant, finally get my cell phone chip problem resolved (about $15), and buy an orange Fanta before commencing the long hike back up the hill. Initially, I had planned to take a cab for about a $3 fare, but I decided that the hike would do me good. It was about an hour's hike up the mountain, and when I found my legs sore the following morning, I almost panicked at the thought of adding sixty or so pounds to my back and climbing ten thousand vertical feet over rocky and painfully difficult terrain! I did not have time to contemplate my dilemma for long, however, as that afternoon, Saim arrived and announced, "Tomorrow, we go."

As I began packing and repacking my backpack, trying to eliminate as much weight as possible, I was pleased to hear that we would have four horses to carry our gear some of the way. Because of that, I decided to keep my camp stove, fuel, and freeze-dried food. The Kurds were planning the meals, but in case Richard and I went alone to certain areas, it would be nice to have a hot meal provision along. As it turned out, it was excess weight I could have done without.

The following afternoon, we packed all gear on top of the luggage rack of Saim's Russian Lada (still holding together, but looking a bit worse for the wear after two years of abuse). We then drove out of town for several miles and eventually turned on to a small dirt road leading into the foothills. After a few miles, the "road" became little better than a rough trail and it became readily apparent that the sparsely scattered Kurdish shepherds we encountered were not accustomed to visitors, and none too pleased with the invasion of their territory. Some would come toward the road—at least one holding a shotgun—and angrily gesture for us to turn around. It would not be the place to be without a Kurdish "guide."

We were brought up to this point by an almost "chance" encounter that Richard had had with Saim in Istanbul the previous spring. He had flown to meet Saim and discuss a "plan" for the summer, and during the process he went over Ed Davis's account of seeing the ark in 1943. Richard mentioned in passing that it was a man by the name of "Abas Abas" who took Ed there. Saim had declared, "I know this family," and thus was this year's expedition born.

The main details of the Ed Davis account are well known in ark research circles and have already been summarized previously (see chapter "The Biblical Account of the Flood"). Most researchers are convinced that Ed was telling the truth about what he experienced, but not all agree that Ed was on Mt. Ararat in eastern Turkey. In his defense, Ed never specifically claimed that he knew he was on Ararat in Turkey. He simply described, as best he could remember, the route that he took and the difficulties of getting there.

Though many specific details recounted by Davis seem to reference Mt. Ararat to the exclusion of all other candidates, on the other hand, some of the details shared by Davis seem hard to reconcile with the Ararat of Turkey. For one thing, he mentioned seeing very large, very old, grapevines in or near one of the villages he was taken to. To my knowledge, no one has seen any such thing around the mountain in recent memory. He also mentioned noticing the very strong smell of sulfur at some point of his climb, and he said that some of his "guides" told him that from the peak of the mountain you could sometimes see the glow of the lights of Tehran. This has, for some, indicated he could not have been on Turkey's Ararat because Tehran is over four hundred miles away as the crow flies. Whether we have ultimately found the final location of the ark or not, I believe it is possible that we have found the answers to all the remaining questions of the Ed Davis account—and they definitively point to Mt. Ararat of eastern Turkey as the resting place of the ark.

Many miles and at least an hour after leaving Dogubayazit, we arrived in a small "village" occupied by Kurdish shepherds. (A village, in this case, consists of a few stone houses loosely scattered, maybe a tent or two, and some animals.) From the curious gathering that soon surrounded us, it was clear that they were unused to visitors from a foreign country—or anywhere for that matter. Saim later told me that we were the only non-Kurdish visitors (other than the military) that they had ever had (except one of the family also told us that a surprise guest had dropped by their village many years before).

We were summarily invited into one of the homes for the obligatory welcome "chai" (tea) while we awaited the arrival of the horses and guides. The small, stone house we were ushered into was about twelve feet by fifteen feet square with logs supporting a thatched roof. The furnishings consisted of one wooden chest containing necessary foodstuffs and household supplies, and in one corner, the sleeping mats and pillows they used for sleeping when necessary. There were a couple of children's backpacks and miscellaneous items hanging from the walls, and one window for both air and light. We learned that eleven family members lived in this one tiny home: the parents, one set of grandparents, and seven children. According to Saim, at least three times in past years, the military had destroyed this village and tried to drive these Kurdish shepherds off the mountain and away from their traditional homeland and lifestyle. But they are a determined people and for many of them, this is the only life they know or want.

As we were drinking chai, we used the time to question them about their life and about the story of Noah's Ark. I asked where the children go to school and was told, sadly, that there was no longer any school. They seemed to brighten when they learned that I was a teacher and they explained that they once had a school, but the military had "stopped" it. We later saw the burned-out wreckage of the school building that people were now beginning to utilize as living quarters. We had also seen electrical lines leading to the village so we asked "where is electricity?" but were told that the military had cut off their power as well. At one point, Richard asked, "Where is bathroom?" and when Saim translated, they laughed. "Everywhere is bathroom. Anywhere is bathroom." They seemed amused that we would want to know. As I've said, hygiene is hard to come by.

Gradually, the questions turned to Noah's Ark and it was clear that not all were in agreement about welcoming our presence and reason for being there. Mehmet explained that his grandfather had been to the ark many times. As I was trying to resolve how these Kurdish shepherds from eastern Turkey might have encountered Ed Davis in Iran in World War II, I asked if their grandfather had been to Iran during that time period. Mehmet became animated as he explained, "Yes, yes, our main village is in Iran." The Abas family actually covers a large area of territory from northwest Iran into eastern Turkey and the south portion of Mt. Ararat. Their main holdings and village are right across the border in Iran. They see this entire area as part of their holdings and homeland—national borders are not a part of their invention or understanding. Thus, one "problem" with Ed Davis's account was seemingly resolved.

Mehmet went on to explain that his own father had also been to see the ark, but he had died ten years prior without taking his sons there. In fact, about two years before he died, he had taken Mehmet up the mountain and pointed out the area where the ark rested. He asked Mehmet to promise him that he would never go there. His father explained that if the ark was "found" again, the world would end. There were some in the Abas village that were angry with Mehmet for even talking to us, let alone agreeing to help us climb the mountain. He also claimed that about fifteen years previously, a climber had come down from their mountain and tried to ask them about Noah's Ark. His father had received this man and given him chai, but told him in no uncertain terms, "No Noah's Ark. Go!" That was apparently the last time the question arose from an outsider.

I then asked him if there had ever been grapevines or vineyards in this area. At first Saim had difficulty understanding "vineyards," but when I explained that grapes were used to make wine, he understood and translated the question for the shepherds. They nodded. "Yes, yes, but no more." Many, many years before there were vineyards, but they had since been cut down and destroyed. My impression was that such destruction was performed by the military. None still existed to their knowledge, yet this was still another positive and persuasive confirmation of Ed Davis' recounting of seeing large, old vineyards near the village to which he was taken. "Were we in the very village that Ed Davis started from?" I wondered. The area that we were in was fed by mountain streams and appeared very fertile compared to other areas of the mountain I have seen. Grapevines were certainly possible and even likely in this area.

Later in our trip, Saim spontaneously shared with me that during one of his forays into the Ahora Gorge this year, he had climbed down to a cave entrance (which he initially mistook for a structure of some sort). It turned out to be a very large cave whose floor was covered in ice for the first thirty feet or so. But inside, he told me, there was a terrible smell. Intrigued, I asked if the smell could have been sulfur, but he did not know or understand the word. He did, however, agree that it could be similar to the smell of rotten eggs—"Yes," he said, "rotten food." "Sulfur for sure," I thought. If so, that would mean another specific detail of Ed's account verified of Mt. Ararat in eastern Turkey. Saim could have been in one of the very caves that Ed and his guides stayed in.

Mehmet agreed that he would supply horses and help us up the mountain, but he would not go with us when we

The 2003 Search For Noah's Ark

continued on to look for the ark. He seemed very afraid of breaking his promise to his father and he was worried about the consequences of finding the ark—both for him and his village, and for the world. Three Kurds, Mehmet, Zachee, and Mehmet Salih, would act as our porters and horse-handlers and we would have four horses. One other man, Hasheem, agreed to meet us with his horse the following morning, even though he was unhappy with the object of our quest.

When the horses arrived and the gear began to be loaded into saddlebags, Saim, Richard, and I started out walking. It was a beautiful afternoon, and the rolling, grass-covered hills betokened a pleasant hike ahead. I tried not to look up at the imposing peak of Ararat, nor to think about the three days of difficult and potentially painful climbing that lay ahead. It was about an hour before the horses caught up to us, and when they did, I was in for a pleasant surprise. One horse was carrying the bulk of our gear—three backpacks. Two of the remaining horses, though saddled with bags of gear and food, were designated for Saim and me to ride! The last horse was saddled for Richard. That was good news beyond anything that I had anticipated, and I am sure my countenance brightened considerably at the prospects of *riding* up Ararat!

As we made our way up the mountain, we would variously encounter sheep, cattle, and shepherds' camps where elements of the extended family lived. Twice on the first afternoon we made the always-obligatory stop for chai at one of the high camps—usually a couple of stone houses, but sometimes only a tent. At the first stop we were introduced by Mehmet to a young man who walked with difficulty and in much pain. They tried to explain that he seemed to be experiencing some progressive disease that was slowly robbing him of mobility. They asked if we had anything that could help. Richard had an extra bottle of aspirin, so he dug it out of his pack, and then I asked if we could pray. They seemed grateful when Richard and I laid hands on the young man and prayed for his healing, so they went and found a girl who also had some sort of physical/mental limitations—apparently from birth defects. We also prayed for her and I found a small bottle of Excedrin from my pack to leave for her. Though I was hoping that we would come back through their "camp" to check on our patients' progress, it was not to be.

The first afternoon of climbing, hiking, and riding did not actually get started until after 3:00, but as I was still anticipating an all-night climb, I was grateful for the early provision of the horse. Many times I recall thinking from the back of the horse, "Now, this is the way to climb a mountain!" I was continually astonished at the horse's ability to carry me as well as the extra gear up a precipitous mountain path over rocks and terrain that I would have had difficulty negotiating on foot and without luggage! Very quickly I learned to say "good horse" in Kurdish: "rende hesp"!

As darkness was setting in, we arrived at a grassy plateau with another shepherds' camp nearby and our guides proceeded to unload and set up camp. Again it was a pleasant surprise for me because I had been resigned to the task of climbing all night. I found my pack and set up my tent while Saim proceeded to make supper. Basically it was potatoes and tomatoes cut up and boiled together, but after we'd ridden and climbed for several hours, it seemed like a feast. I used my headlamp to help navigate in the dark and complete the setting up of my tent. Though I have always marveled at the Kurds' ability to negotiate steep, rocky, perilous mountain trails in pitch-darkness with absolutely no light, Mehmet seemed to appreciate the benefits of my headlamp, and that first night, he utilized it more than I did. I wondered, "Hmmm, has progress come to the Kurdish shepherds?"

The following morning, after breakfast of bread, honey, and chai, I packed my bags and then waited as the horses were gathered, saddled, and packed. While we waited, Saim, Richard, and I meandered down to see the shepherds' "home" (basically a rock wall with a tent over it). We were summarily invited to join for more chai. The shepherds live a very simple life, providing most of their own needs from the things around them: wool for clothes and blankets and rugs from the sheep and goats; milk from same; eggs and meat from the chickens, water from the streams. But, to most of us, it would also seem a very difficult, lonely, and monumentally boring way of life. To them, it is simply life. They know no other. For most of the children we encountered at the high camps, we were the first non-Kurdish "strangers" they had ever seen. They were very shy and unfriendly—even surly at times. At this last high camp, one little boy (maybe seven years old) had broken his shoulder and it had not been medically treated and consequently had never properly healed. His mother made him take his shirt off so that we could see the "defect." Not once did he smile or say a word. Once again, Richard and I asked if we could pray for him. Though his mother seemed grateful, the little boy pulled angrily away when we were finished.

We headed out of camp—riding—at about 8:00 a.m. and already I was amazed that my legs could be sore without having hiked any appreciable distance and without having carried my pack at all. Part of it was undoubtedly the result of clinging to the horse's back with my knees and thigh muscles in a position that I had not had opportunity to utilize regularly for almost thirty years previously. The rest was exacerbated by the fact that my legs had already been sore from my hike to town the day before.

The trail on this second day became much more difficult, necessitating that many times I dismount and lead the horse. It sounds like a simple thing to do, but many times I prayed that no one was watching as I struggled and fumbled my way down. "Indiana Jones, I'm not!" I thought. In fairness to myself, I was riding a horse that was saddled with bulging saddlebags placed over the saddle and on either side of the horse. That meant that my legs had to extend in front of the heavy bags in order to fit my feet in the stirrups. It was an awkward position and unfortunately, I could not pull my foot backward from the stirrup to extricate it at will, because it was pinned by the saddlebags. In order to dismount, I would have to lean far over and with my right hand release my right foot from its stirrup, and then do the same with the left. I would then swing my right leg up over the horse's neck so that I was sitting sidesaddle, and then quickly pirouette as I slid down in front of the saddlebag.

Sometimes the decision to vacate the horse had to be made quickly as he (she?) struggled up a steep, narrow, rocky "trail"—and, if I were to jump at the wrong instant, my landing could potentially propel me over the edge of some precipice (and result in great discomfort, to say the least!).

Before I "perfected" this technique, at least once, I tried to dismount with my left foot still in the stirrup (normal style). We were going up a steep section of trail, and my foot remained trapped between the saddlebags and the now-twisted stirrup. I had to let go of the rein in order to keep my balance (actually, the horses were not bridled—they had only a halter and a piece of rope or light chain attached to one side to be utilized as either a tether or a "rein" depending on what was needed) requiring me to hop awkwardly alongside my horse before I could convince it to stop and extricate my foot. The Kurds following seemed helpful and sympathetic but I was convinced that, inside, they enjoyed a hearty chuckle.

We made camp fairly early on the second day—perhaps around 4:00 p.m.—because we had arrived at the end of the line for the horses. We were in a nice, level, grassy area with a stream running through one side, and plenty of grazing for the horses. Looking up at the huge and treacherous pile of volcanic rock behind me gave me no pleasure at the thought of the coming day.

That night, Richard apparently had trouble sleeping—usually he is out within five minutes of going prone—and I actually slept through most of the night for the first time. It was he or I, but not both, who slept on a given night. As I went to bed, I consoled myself with the thought that maybe I would not have to carry my full pack to high camp on the morrow. I was wrong, but luckily I did not know that at the time, or I might not have slept so peacefully.

The 2003 Search For Noah's Ark

The following day was brutal. Again we started with the same light breakfast—bread, a choice or two of spread, and chai—packed our things and headed up the lava-field. It did not take long for the thought to come to me, "You've done this before. And you swore that you would never do it again! What, exactly, is wrong with you!" There is simply no way to put into words the agony you feel when, nearing the brink of exhaustion, you come to the "edge" that has been your goal, seemingly forever, only to find that the hill goes ever on and you are apparently no closer to the top. My only ongoing consolation was that this year, though it was a very difficult climb, it only lasted for roughly four hours (instead of nine or ten as in 2001).

We arrived at about 14,000 feet to find a campsite in the rocks. Saim told me it had variously been occupied in the past by the PKK (Kurdish "terrorist" organization), and the military. It basically consisted of places where rocks had been removed and even piled into walled enclosures, and the ground in between leveled. A couple of the areas were large enough to set up a tent, so we made camp there.

By the time I had straggled into camp, the Kurds had already set up the other tent and were in the process of making dinner, even though it was barely 1:00 p.m. I was somewhat chagrined at having taken so long, but Mehmet Salih, the youngest of the helpers at nineteen, made a muscle pose and pointed at me, "You, Sylvester Stallone." I was not sure if he was truly complimenting me in deference to the weight of my pack, or if he was making his own private joke at my expense. "If," I thought, "I'm Sylvester Stallone, it must be after being hit by a train." Outwardly I said, "No, me Mickey Mouse." Serious or not, his comment helped alleviate a bit of the embarrassment of being so slow, and I proceeded to find a spot to set up my tent.

We had come to the edge of the Abas territory and Mehmet would go no farther with us. He pointed us up and to the right, and indicated that the ark area was "somewhere up there." Our goal was in the area of between fourteen and fifteen thousand feet of altitude, so, vertically, we did not have a long way to go, and we were planning to use this camp as our base for several days if need be.

After lunch (which was also dinner—our one solid meal each day came whenever we set up camp regardless of when that was), which was some kind of vegetable, rice, noodle concoction, the weather turned and it began to hail. Clouds rolled in, visibility diminished, and most of us lay down to rest. Mehmet Abas and Mehmet Salih headed back down to the horses with the promise to return early the next day. After a bit of fitful dozing, I was bored stiff. There was nowhere to go, nothing to do, and I was rested and ready to go. The weather did not agree. When it finally cleared, there was little we could accomplish in the few remaining daylight hours, so I went to bed. Tragically, for me, after lying around all afternoon, I could not sleep as my tent mate snored happily away. Perhaps the only "bears" on Mt. Ararat are the people who inhabit it! In any event, it was a very long night.

That night, before settling into my tent for the duration, I believe I had an epiphany as to another aspect of Ed Davis's account. He mentioned that he had been told by his guides that from the peak of Ararat, on a clear night, one could see the glow of the lights of Tehran. Many have questioned how it might be possible to see the lights of a city over four hundred miles away. Some have suggested that he (or they) must have meant Yerevan, which is a large city across the border in Armenia. Others, including author Bob Cornuke, have utilized the "lights of Tehran" claim (among other things) to move the whole mountain closer to Tehran by finding another candidate for Ararat. That evening, however, I could look down from where we were camped and easily see the lights of several small cities in Iran. Individual lights were clearly visible. All of a sudden, I realized that Yerevan was so close that it was easily visible across the border into Armenia. There would be no reason for them to tell Ed that they could see the glow from the lights of Yerevan—the whole city was clearly visible. Besides, being from Iran, they would have

The 2003 Search For Noah's Ark

no reason to brag that from the peak of Ararat they could see all the way to Yerevan. Yerevan is not even in their nation. When I remembered that Mehmet told me that their main village was in Iran, it made sense. If they were going to brag about seeing the glow of lights from a distant city, it would have to be Tehran! They could not see the actual city lights of Tehran, like they could Yerevan, but on a clear night, they could see the glow in the sky of the lights from there. Even if they were wrong about where the glow was actually coming from, it was natural that they would brag about being able to see all the way to Tehran. If they were on Mt. Sabalon in Iran, that would not be true. It would not be something to brag about, it would be easily done—they would be able to look down and see the city itself. In my mind, yet another of the troubling aspects of the Davis account was resolved.

Finally, on the fourth day, we were set to begin the real search. After the now traditional breakfast, but with increasingly dry and stale, crumbling bread, we headed up the hill into the area designated by Mehmet as the "ark area." I got up to a peak of rocks and took a GPS reading—about 14,900 feet—and a few photos as I waited for Richard, Saim and Zachee (Zachee was hired by Saim as a porter and was not part of the Abas clan and, therefore, not afraid to go up with us). As I looked for the others, I realized there was a large ice-field that I would have to cross in order to continue. I was not wearing my hiking boots as this was intended to be just our initial reconnaissance mission, nor did I have my crampons, so I had to descend to meet the others and go around the ice.

When we made it to the other ridge and looked down, my heart almost stopped. Excitement arose as I realized that here, indeed, was a very large valley or gorge—complete with streams and even a waterfall or two—very much like the terrain and conditions described by Ed Davis. I sat down nervously with my binoculars to scan the terrain for any rectangular or unusual shapes. The initial excitement began to wane as nothing immediately came into focus. However, somewhat below me, at the base of

Dinosaurs on the Ark

a waterfall that ran down through the valley, there appeared a large, dark object. Nervously, I trained the binoculars on it, but through the magnified view from the binoculars, it appeared to be primarily ice. It did not occur to me at the time to wonder why the ice was black, especially in a canyon of red rock. Nor did it occur to me that it might be a black object covered in ice. I was disappointed, but nevertheless made a mental note to make my way down there and get a closer look.

Before we could get started, as we were all looking at various aspects of the gorge in front of us, a low rumbling began and very quickly escalated into an awesome and even frightening roar. It continued unabated for what seemed to be the better portion of a minute. It must have been an enormous avalanche, but it was not in our valley and could only have been in the Ahora Gorge. Nevertheless, it served as a ready reminder of the constant threat we were under should an avalanche happen in our area. I distinctly remember thinking, "Dear Lord, thank you for keeping that from us." It was not until we had later returned to camp that I realized how much it had unnerved the two Kurdish guides left in camp. Mehmet's face registered immediate relief when he saw me heading back into camp, and he came to give me a welcoming hug. Saim let us know that when Mehmet had heard the enormous avalanche, he had believed that perhaps God was punishing us for going too near the ark. He then confided that several years before, he had decided to break his promise to his father because he really wanted to see the ark. When he got close to the area, a great storm came up and they fled. To him, it seemed that the storm even followed them wherever they went. He was sure that God was angry with him, and they had never dared to try again.

In the meantime, after the roar of the avalanche subsided, Richard and Saim decided that we could cover more territory and see more things if we split up. Zachee was young (twenty-three), strong, and fast, so for some odd reason (the odd couple?), Saim thought I should go with him up over the canyon and down the east ridge of the

The 2003 Search For Noah's Ark

Ahora Gorge. I objected because I was unprepared—I had no boots or crampons or harness—and because I did not want to be traversing dangerous ground with someone who spoke no English. Saim then agreed to go with Zachee while I stayed with Richard. We had two areas below us that we wanted to look in to, so Richard (or Saim?) suggested we separate and each examine one area before meeting again at the bottom and working our way back to camp.

Richard is absolutely tenacious, and the word "perseverance" could be defined in the dictionary simply by putting his picture next to the word. But he is also no longer young. At fifty-eight (in 2003), he was approaching mandatory retirement age from his airline pilot's job at age sixty. His joints do not move as they once did—climbing onto the horse proved to be a task requiring the help of our guides. Maybe, if he had not been with me—and trying to take a second path by himself—I would have made my way down to the object of my interest regardless of the weather. Maybe it was spiritual warfare and the enemy was using doubt and fear to assail me. Maybe it was something else. Maybe I should have stopped and prayed more fervently for direction. Maybe I will never know. But, somehow, when the clouds and fog rolled in and obscured the area below completely, I hesitated. I decided to sit and wait and see if the clouds would clear. I waited ... and waited. And the clouds got thicker. Finally, without going down to definitively rule out this "object," I made the decision to find Richard and make the return to camp. "Besides," I figured, "there's always tomorrow." Famous last words—and I've said them before on Ararat, only to find them untrue. There was no "tomorrow" on this trip either.

I managed to locate Richard and when we eventually made our way back to camp, Saim was already there and very winded. He claimed that they had been to the east ridge and partway down, seen nothing, found the terrain to be frighteningly treacherous, and returned (though, of course, we had no way of confirming that they had actually gone anywhere other than around the nearest rock for a siesta). Saim then claimed that he had gotten a call from the military in Aralik. A spotter in Igdir had seen him on the east ridge (even by telescope, no one could have known it was him!) and called the commander in Aralik. The military hierarchy was furious, according to Saim, and he said that we must go down at once. Maybe that was the case. Maybe Saim was just so exhausted from his trek that he did not want to consider the prospects of another day of doing the same. Maybe, if I had not been so tired myself, I would have protested more vociferously. Or maybe I would have just stayed myself regardless of what he did. Maybe I'll never know those things either, but somehow the decision was made to strike camp and head back down. I knew that the first stretch of the trek would be the most difficult, but I took some consolation from the fact that I knew the horses would be waiting below, somewhere around twelve thousand feet of altitude. I only had to carry my gear that far.

Coming down is usually preferable to going up, especially when carrying a lot of weight. But at 48 years old, and after my having endured three or four knee surgeries from athletic injuries, the arthritis in my knees can make the downhill climb excruciating. In addition, wearing my heavy boots each year in the past has caused me to lose the big toenail on my right foot. So, this time, even though I could have used the ankle support, I opted to wear my sneakers on the way down. It was a good choice.

I think (but it is a bit of a blur in my memory banks) the climb down to horse-level took about two hours—it only seemed like eight. But once I unloaded my pack and laid it in the grass, I do remember hoping against hope that we were just going to set up camp there.

We did not. The Kurds quickly rounded up the horses, saddled them, packed the bags, loaded them up, and started down. I was again handed the chain of "my horse." It had been given to me to be the handler and rider of one particular horse. I did not think much of it at the time, but later I was to find that it was a bit of an honor that they entrusted their horse to me. All along I assumed that they were quietly laughing at my foibles mounting and dismounting the given horse. But that evening, back in camp, I noticed the youngest of our Kurdish guides talking to his friends and pointing at me. I figured that I was the butt of some humorous story about fumbling down the mountain and needing to sometimes use a rock in order to mount my horse over the saddlebags with my stiff and sore knees. However, Saim leaned over and confided to me that Mehmet was telling his friends how I had handled my own horse all the way up and down the mountain. Hopefully, Saim was not merely lying to make me feel better, but if so, it worked nonetheless.

After the horses were loaded we headed for "low" camp. I'm not sure whether my horse was handed to me, or me to him. But the terrain quickly became too difficult to ride and I tried desperately to find a graceful way to dismount. Dismounting proved to be an even greater challenge now that my legs were tired and sore. Climbing on the horse would bring some initial relief, but the longer I sat, the stiffer I got. Sometimes just unbending my knees was excruciating, and the thought of jumping to the ground was painful in itself. When I did hit the ground, it would take several stumbling steps before some hint of flexibility would begin to return.

For the first couple of hours, the walking and riding were probably fairly evenly divided. As dusk began to approach, I suddenly realized they had no intention of stopping for the night until we were back at the starting point. In the morning, we had started our climb at 8:00 a.m., and it was now getting dark at 7:00 p.m. and there was no sign of letting up. Somewhere during that process, I made the very sober judgment that I was through. "I've gone as far as I physically can; I've done as much as I am able," I thought. I knew that I had made such decisions or statements before—several times as a matter of fact: "Never again will I come here in search of Noah's Ark." God should have chosen a younger, more physically able man. In the past, though, it had struck me as more of an emotional reaction to the continual sense of failure rather than a reasoned response to my physical abilities. This time it was almost an epiphany: "I simply physically cannot do this anymore—maybe with two good knees, but not as I am." It was almost a relief to realize that it would have to be someone else who carried on the task—maybe by using the information that we have discovered.

The last three hours on the horse were in pitch-black darkness. How the horse could see anything I do not know, but most of the time I felt foolish trying to lead him. Sometimes I could barely make out the silhouette of the rider in front of me, but I could see nothing of the trail. It continued to astound me that the horse could carry rider and baggage down a steep and sometimes treacherous trail, and now in complete darkness! But I also knew that I could not walk the same trail in my condition. As painful as my knees now were—in my position on the horse, I could not flex or reposition them to bring relief—as I began bemoaning my position and complaining to God, it came quietly into my mind, "Would you rather be walking?" I remembered coming down the mountain by night in 2001 and I realized how much better off I was on the horse than on foot. I began thanking God for my "rende hesp" and doing my best to simply ignore my knees.

Mehmet had long ago requested the headlamp from my pack, so the myth of the Kurdish shepherd immune to the dark had been shattered and I could see the bobbing light some distance ahead. Even with him leading the way by using my headlamp, several times we lost the trail and had to backtrack.

It was somewhat after 10:00 p.m. when we stumbled back into the initial village and were soon surrounded by the curious and the children who were still awake. Even sitting on the ground was now painful, and not even the obligatory cup of chai offered any solace to me now. After unloading all our gear, we had to await the arrival of the "Muratmobile" (the faithful Russian Lada). I decided to award Mehmet my headlamp, so I showed him the inner workings and gave him three extra batteries. He seemed inordinately pleased and I suspect that the day of the "lightless" Kurd will soon be over—civilization has come to the mountain!

Though the Lada soon arrived, and we quickly loaded our gear, the drive back to town took us over an hour. Funny how interminably long a simple hour can be. My first shower in five days was with warm water, and I doubt seriously that anything could have felt better.

Though I should have been exhausted and slept like a rock, all sorts of questions plagued my sleepless night—not helped perhaps by the return of one of the bears of Ararat! Why had I come again? What had I actually accomplished after all these years of trying? Have I really been responding to God's call on my life? It seems that, perhaps, I have learned some crucial things, but I do not even know how to share them without endangering the lives of the Kurds involved. Was it the Lord who prevented me from going down to examine the object of interest in the gorge? Or was it the enemy trying to thwart me? In hindsight, had I known that I would not be able to return to that area, I certainly would have gone down for a definitive look.

The following day, Richard had many of the same nagging doubts and questions. The feeling that, perhaps, we had come close, was impossible to shake. I was not positive that we had come close to anything other than ice, but I now wondered. Prior to this trip, I had been convinced that the ark had to be somewhere in the Ahora Gorge, and it had to be mostly buried—if only in snow or ice. Maybe if I had not been so convinced of this, I would have made more effort to make the climb down and be sure. Maybe ...

In any case, I had both a camera with slide film and a pair of binoculars that doubled as a digital camera. Through the binoculars, I had taken seven pictures of the region we were in, and I had taken two specifically of the dark object of interest. "At least I have something to go home with," I thought. "It will be interesting when I download the digital photos and get to study this object more closely."

As it turned out, it was indeed interesting, but not for the reasons I had hoped. I had taken seven pictures through the binoculars, but only two of this particular object. When I eventually made it home and downloaded the pictures, five of the pictures were fine and clear. Only the two pictures that should have shown the "object" were blank. It is clear that I did take seven pictures, because the camera records it as such, and seven are downloaded onto my computer. It is just that two of the pictures are completely black—the two that were of the unusual object in the gorge!

That afternoon I made the decision to leave Dogubayazit several days early and try to catch standby flights back to the States. The thought of hanging around eastern Turkey with nothing more to do was almost enough to drive me to drink. But first, I resolved to find my friend of previous years, Salih. I had not seen him in two years and we had a nice time catching up as he drove me outside of town to visit some mineral springs. He invited me to be his houseguest for the evening and the discussion inevitably turned to Ararat and Noah's Ark.

When I initially asked him if he could get me a permit and help me to climb Mt. Ararat, he said, "Of course." But, when I asked him if he believed that Noah's Ark was somewhere up there he simply said, "No." Yet, without any prompting, he began telling me of an Armenian man he had met the previous spring. They traveled around together for the day and then went back to the motel in the evening for a cool drink. The man invited Salih to his room to show him some things of interest. One of the things he had was a photo of Mt. Ararat, taken from the Armenian side. He told Salih that eastern Turkey and Mt. Ararat used to be "Armenia" before the Turks drove them out and killed many Armenians. On the photo, he showed Salih two pieces of what he claimed was Noah's Ark.

As Salih began to explain, and draw for me, I realized how similar to Ed Davis's account this was. The Armenian said there were two pieces of the ark and the upper piece was largely covered by the mountain. Part of the end was purportedly sticking out, and the lower piece was mostly exposed when there was no snow. Salih said the two pieces that he saw were in a "valley" and my mind immediately turned to the Ahora Gorge. "You mean in the Ahora Gorge?" I asked. "No," he said, "Not there." And he then drew a picture, locating the valley to the south side of the gorge, seemingly exactly where Richard and I had been the day before! My heart almost stopped. And now the questions of the day before took on added significance. I realized that the issue of making my way down to that black patch of "ice" had not seemed as essential because I just did not believe that the ark was there. What is the old saying? "I would not have seen it if I had not believed it." Salih gave me the name and the phone number of the Armenian man so that I could call him from the States. Suddenly, the decision to retire from actively climbing the mountain in search of Noah's Ark—sober or not—dissolved in a heartbeat. How close had we been?

Endnotes

1 Cornuke, Robert, & David Halbrook. *In Search of the Lost Mountains of Noah*. Nashville, Tennessee: Broadman & Holman Publishers, 2001.

Chapter 17

THE 2005 SEARCH FOR NOAH'S ARK

In September of 2005, I flew to Istanbul via JFK and then Paris, and although the flights were essentially on time and uneventful, it still seemed an interminable journey. Richard Bright met me at the Istanbul Airport and we traveled by cab to his favorite nearby haunt, the Airport Inn. It is a pleasant enough locale and conveniently close to the airport, but becoming more well known and therefore more expensive (this year around $75). My room was on the fourth floor and had very pleasant views overlooking the bay. I showered, changed, and tried to nap for a couple of hours before meeting back with Richard. The attempts at sleep were mostly ineffectual but it was nice to at least recline and listen to soft music after sixteen hours on planes and many more in airports.

At supper (steak and French fries—a first for me in Turkey!) we discussed a number of options and then retired to Richard's room where we pulled out the photos, written accounts and books and began to try and piece them all together. Attempting to do so makes one realize just how discordant all of the various pieces are, and how difficult it is to weave them all together into one coherent story. We came up with some options, but no definite conclusions or plan. Next morning (after a great breakfast of cheese omelettes and watermelon and various breads and cakes), we flew to Erzurum and were met by Saim. We then drove to a trekking office (where we also met Ahmet Ozturk, a Turkish mountaineering guide) and discussed our various options there. Richard's primary concern was getting into the Ahora Gorge to check out an area in the Araxes Glacier where an "eyewitness" had claimed to have seen the ark some forty years ago. Last year, Richard

The 2005 Search For Noah's Ark

and Saim had ventured into this area, but were unsuccessful in locating anything definite.

On the other hand, I had traveled to Turkey and Armenia in the spring of 2005 in a largely unsuccessful effort to retrieve a purported ancient map and photo of the ark's location from an Armenian farmer. Though I was unable to locate him, I had received a written portion of the account from my friend Salih. In my last telephone conversation with him before leaving for Turkey, Salih had assured me that he now also had a copy of the map and he would preserve at least one copy for me. The written account, when translated into English, had some intriguing parallels to the Ed Davis account, and I was hoping to put the written account together with the map and follow up those leads on the mountain. Saim had basically concluded that the whole story of an Armenian map was a scam concocted by Salih or one of his associates and not worth wasting time on. He clearly did not want me to contact Salih before we had the chance to climb (though the map would be of little value after we came down from the mountain). My hopes were pinned partially on the map being able to eliminate the seeming inconsistencies of the written account (it mentioned the location of the ark as being on the "south side" of Ararat, but it also placed it near "Lake Kop" whose location is on the north side of Ararat).

We ended up eating supper in Erzurum without reaching any consensus and not leaving for the mountain until after 6 p.m. The drive seemed to be the longest part of the journey so far—the road was under construction, and we did not arrive at my "favorite campground" until after 11:00 p.m.

For most of the next two days Richard and I were largely confined as prisoners in our room at the campground. Even our meals were brought to us and we remained out of sight. I used the time to rest and I read three of the books I had brought, but a person can take only so much "rest" when confined to quarters. We made the mistake of venturing "out" to breakfast at the campground restaurant on the first morning because the campground was basically empty and we were the only two people up for breakfast. Unfortunately, a cab soon dropped off a "tourist" that we assumed to be harmless. He turned out, however, to be part of the "Ark Symposium" that was just ending and he was apparently part of the Italian Palego's team. He came into the restaurant and immediately recognized Richard from the photos on the wall. He came over with a big grin and tried to introduce himself. Fortunately, he got Richard's name wrong (it came out something like "Richard Wight") and Richard simply denied it and walked off. That seemed to confuse the gentleman for a while, but he hung around with his camera for the next hour or so looking for a shot. We returned to seclusion.

Saim kept telling us that Salih and others were part of the local "secret, civilian police" who were informants and if our presence was reported, could make our access to the mountain difficult. He felt that his "two-week" guarantee would be voided if it became public knowledge that we were around and headed for the mountain.

However, all of a sudden, on Sunday, everything changed in a heartbeat without even an explanation as to why. Saim let on that he had told Salih that I was there and it would be fine if I visited with him after all. I got hold of Salih by phone and he dropped by at supper for a brief visit. But I was in for an even greater surprise when, though he had promised to keep a copy for me, he let me know that he had, instead, sold his only copy of the map to a French researcher with whom he had been in negotiation. He then offered to travel with me to Armenia to buy the "original" for only $7,000 (up from $3,500 in the spring). I am not sure why he would promise to keep one copy for me and then renege, but it was a blow to my hopes for additional information, and it probably meant he was looking for any additional angle to sell the information again. If I were a rich man, I might have been tempted to forgo Ararat for another trip to Armenia, but then, even if

I had the money, it seemed too much for too little. And it would have kept me off my favorite mountain.

On Monday morning, Richard and I were all packed and ready to head for the mountain, but Saim never arrived on schedule (of course, "schedule" in Turkey is very loosely interpreted to say the least). The morning began to drag into the afternoon and I was about to seek out lunch when Saim finally arrived with a "problem." He had been directed in town to go to his office because "someone was looking for him." He showed up only to find a member of the "secret police" waiting for him. The man wanted to know what "David Larsen" was doing in town–someone who knew I was in town had obviously passed along the imformation to the police. But who knew I was there? Salih? I am sure he would not have intentionally betrayed me, but someone may have overhead him talking of me. But Salih had left immediately after talking to me in order to go to Lake Van for the evening on business for Saim, and he was not even back in town yet. In any case, we now had to wait for this "policeman" to show up and interrogate me.

When he arrived, he wanted my passport and a picture, but Saim went off with him to a corner table in the restaurant. Somehow, Saim managed to take my place in the "interrogation" and he refused to allow me to be questioned (since I was a "tourist" there as his guest). I think Saim was worried that I would not lie and I would tell too much of what was going on. Nevertheless, the man left with dire warnings to Saim that he was now being "watched," and he would need to be "very careful." Shortly after that however, we packed our gear on top of the old, dilapidated, Russian Lada and headed out of town in broad daylight toward the mountain. Anyone "watching" could not have failed to see!

September 12, day one

Because of the "secret police" delay, we ended up not beginning our climb until roughly 2:00 p.m., when we met up with the horses on the mountain. It is almost pleasant being able to start off hiking in sneakers and carrying no gear except for water. Unfortunately, the horses, welcome as they are, can only help out on what amounts to the easiest part of the mountain anyhow. Nevertheless, I never complain when someone—or something—else is carrying my gear up Ararat. We ended up hiking/climbing for about four hours, and since the horses and handlers went by a slightly different route, they had already arrived and set up camp by the time we got there. Supper consisted of some kind of potato soup—our only meal since breakfast many long hours before. One of the Kurdish shepherds helping us was Ahmet Abas, brother of Mehmet, who had helped us two years ago, and purportedly a member of the Abas clan who befriended Ed Davis in 1943. Ahmet Ozturk, the Turkish mountaineer who maintained a trekking business in Erzurum, it turned out, had been with Jim Irwin in both 1985 and 1986 (and thus with Richard as well, but neither remembered the other). Two other shepherds, Mehmet and Halik made up the remainder of our crew. Camp was set up on a grassy knoll at about 10,500 feet of elevation, and it was pleasant enough, but I spent another long and sleepless night on Ararat. I was in a tent by myself this year since the other tents were larger and all got shared, and even though it was neither cold nor overly unpleasant, I simply tossed and turned all night. Stress? Ararat seems to have the ability to invade my sleep, and after a while it can be quite painful to go so long without any. I used to blame it almost entirely on the snoring of various tent mates, but this year I was in a tent entirely on my own, and I still could not sleep.

September 13, day two

In the "morning," the Kurds were up at 4:00, talking loudly and making chai (tea) and smoking cigarettes. Why they were up at that hour, I will probably never know, but they made no attempt to whisper or even to talk quietly. After that, any further attempts at sleep were fruitless. I finally got up around 6:00 and fiddled around with keeping my journal. "Breakfast"—dry bread and a hunk of cheese—was at about 7:30 and then we packed and headed off around 8:30. I was heartened to see the sun rise above the rocks in

The 2005 Search For Noah's Ark

exactly the location that I had predicted the night before. Dogubayazit has always been referred to as the "south" side of the mountain, but I have maintained that it is actually closer to southwest. Both Saim and Richard had predicted different sunrise positions, but they were significantly off target. We were now actually on the true south side of the mountain, so if the Armenian account had any validity, we were nearing at least one target area of our search.

It is a very annoying feature of ascending Ararat that, because of the steepness of the route, you can only observe one distant ridge at a time. Often you will climb toward a ridge that appears to be your final goal—over tumbling volcanic boulders; a steep and exhausting journey—only to find that you are, instead, approaching a small ridge in the much vaster scheme of things. There will either be another ridge farther back and higher up, or another valley to climb down into, another boulder/avalanche field to cross, and then another steep, seemingly endless hill to climb. The adage, "once high, stay high" never seems to pertain to Ararat. And the huge debris fields of volcanic boulders just seem endless.

This particular day entailed about five hours of climbing to reach our camp destination and I ended up arriving just as the horses did (they had taken a more circuitous route and I had not even seen them for the rest of the climb). This time camp was much more rocky, windy, and generally uncomfortable than the day before. And this time, I had to set up my own tent. According to the GPS, our second camp was only at 12,600 feet, but this was as far as the horses would go on this side of the mountain. There did appear to be a "valley" to the north of us that would require a closer look, and then we planned to head up to the "high camp" where Richard and I had stayed two years before—heading to the "ark area" according to the Abas clan. From there, the plan was to search that region thoroughly and then head up and over the ice to the east ridge of the Ahora Gorge. More and more I have wondered if anything is still visible of the ark. Maybe it is by now completely buried by avalanche debris.

As I lay there, I found myself wondering how my legs could be so tired and sore without having even carried my pack anywhere. That did not seem to bode well for the morrow! Richard Bright never ceases to amaze me. He is not exactly in prime physical condition for climbing volcanic rock piles, and this year, he had a debilitating cough that often would not allow him to sleep at night. But he is the Energizer Bunny of ark researchers. He is absolutely tenacious—indefatigable—and every single year since 1998 that I have been with him, he has adamantly insisted that it was his last year—and yet this trip represented his twenty-seventh trip to Turkey! Not all of those trips have included climbs of the mountain, but my guess would be that he has been on the mountain more than any other three researchers combined. Part of him seems to actually enjoy it (a small part!), but I have known him to stay in eastern Turkey—in squalid conditions—and put together two or three climbs in one trip! I am completely done in after one climb, and the thought of coming down and then going back up again is almost horrifying. The mental exhaustion alone after coming down empty-handed is enough to demoralize me for at least another year—usually two! I have never even contemplated a second climb on the same trip, even when I had the time or the option, let alone done it. I believe that if this year does not reveal the ark, I will have exhausted my areas of research (and my abilities) and I will surrender the task to younger legs and heartier souls. (And as I was lying in my tent and writing these things in my journal, it began hailing on the tent, and the wind felt as though it would rip the tent away—had I not been in it, the tent would undoubtedly have been long gone. I prayed in desperation that I would get sleep. I didn't.)

After I entered the tent and the hail began in earnest, I decided the better part of valor would be to spend my time writing in my journal, reading my Bible, and praying. I hoped that would occupy me until at least 9:00 p.m. so that I would not fall asleep too soon. One of the worst aspects of my times on Ararat has been my utter inability to get any sleep. Unfortunately, I lasted only until about

6:00 p.m., when I decided to rest my eyes while I prayed a bit. When I finished "resting" my eyes, I woke up and it was 10:00. I actually believe that is the most continuous sleep I have had on Ararat to date, but unfortunately, after that it was another long night of tossing and turning, prayer, and sleeplessness. When I finally succumbed to the call of nature at around 1:00 a.m. and bundled up for the exodus from the tent, the ground was white with hail—and the wind was relentless.

September 14, day three

In the morning, the plan was for Saim and Ahmet (senior) to climb the valley to view one site of "interest" and to look for others. Richard and I, meanwhile, climbed toward the high camp where we had stayed two years before and searched the areas along the way. We traveled light with only binoculars and water—primarily for reconnaissance of any other potential sites of interest. My goal was to get to the higher camp, and it seemed a waste of a good day to hang around where we were. We did not find anything of note, so once we were back in camp, it was just a matter of "preserving energy" until Saim and Ahmet returned with their report. As I sat in camp, preserving energy, I was able to look more thoroughly through the binoculars at the area being investigated by Saim. A good ways up in the ice, there was a large, dark, rectangular area with what appeared to be a straight line under the snow and ice extending away from the opening. It was the first real interesting object I had ever seen on the mountain. It seemed too good to be true that we might have made a discovery only three days into our trip, but we could only sit and await the return of the others.

    I fully expected Saim and Ahmet to come back saying that there was nothing there but a rock or cave, but when they arrived at camp, Saim seemed actually excited. He claimed he was almost 100 percent certain that they had found something. He had digital photographs taken from a distance, but they had not been able to cross the ice without proper gear. (Why they went up there without the necessary gear, I will never know, but that meant we would have to spend another day at this low camp in order to properly investigate the "object.") It was very hard not to get my hopes up even though I made mental notes of all the strikes against the potential object. It was positioned up against a sheer cliff in a place that it seemed would have been impossible to arrive at by falling. If it was part of the ark, where could the other part possibly be? If it was the upper piece, there was no place for a lower piece and it was probably too low for the upper piece (certainly no higher than 13,500 feet). If it was the lower piece, then it seemed equally unlikely that there was an upper piece anywhere in the area that we could see. Still, I had to admit that it warranted a closer look.

The dark, rectangular area in center of photo—with horizontal "roofline" extending back—seemed less and less "arklike" as we drew closer. [Author's photograph, September 2005.]

The 2005 Search For Noah's Ark

September 15, day four

The next morning marked my fourth consecutive day on Ararat, which officially made it my longest excursion yet. It was hard to believe that all my previous trips would already have been finished by this time, and this year we had not really even gotten started yet. I headed off for the "ice object" at around 7:30 a.m. with Richard following somewhere behind. Turkish mountaineer Ahmet carried the rope and Saim followed sometime later with some of Richard's gear. I carried only crampons, ice axe, camera, binoculars, and harness. The rocks on Ararat are seemingly interminable and make for agonizing climbs. It is hard to put into words how painful and never-ending the climbs can seem. Today was a two-thousand-foot elevation gain, but one often slides back one step for every two forward—it only seems that it is two back for every one forward! Climbing loose stones is simply the worst. When I finally arrived at a level across from the ice object, through the binoculars, it began to look less and less like an ark and more and more like a hole. I think that I would have turned back at that point if Richard had not arrived and thought he saw a "beam" protruding from the ice. I was convinced by now that there was no "object," but I realized that after I returned to the tents, we would be done for the day anyhow. The thought of sitting around camp for endless more hours was enough to spur me on. I also decided that it would be best to remove any remaining nagging doubts.

Ahmet had tried to indicate to me that we needed to go farther up in order to "parallel" the object, but as I had by that time already gone farther up and encountered only sheer cliffs down to the glacier, I talked him into coming back to where I was. It appeared to be a rather simple climb down about ten feet of cliff to the valley level below. He, however, removed the rope and anchored it around a rock and used it to lower himself down. I was in the process of making mental comments about how silly it was to use the rope for such a simple climb, when the first major rock he stepped on gave way and plummeted down. The

Author, after the climb from the campsite, preparing to descend and cross the ice-field toward object of interest. [Author's photograph, September 2005.]

next one, the same. If I had tried to climb down before he arrived, I would undoubtedly have been killed or at least badly battered. It was a valuable lesson in caution. When I climbed down, I was dutifully grateful for the rope.

At the bottom, we had to traverse some very steep and rocky terrain to get to the ice and it proved to be more treacherous than it had appeared from above—but that is true of almost everything on Ararat. Once we made it to the edge of the ice, roped together and got our crampons on, Richard had decided (from the ledge above) that it might be too dangerous to cross the ice with boulders cascading down the ice field from above. By now it was getting close to noon, the sun was melting the ice, and an occasional boulder would break loose and come crashing down. By that time, however, I had invested too much time and energy to turn back, so we headed out across the ice.

We worked our way across the steep, and very rough, ice field until we arrived at the "object" roughly at noon. It was a cave—very deep—but a hole in the ground nonetheless. We took a few pictures and tossed some boulders into the abyss to hear how long they took to hit, and then headed down the ice field rather than back the way we had come. There were a few tricky areas where we had to navigate around large crevasses in the ice, but we eventually made it past the end of the ice.

Author, at the entrance to "ice cave" that intrigued us from below. [Author's photograph, September 2005.]

I believe that coming down, especially when tired, is the hardest part for me. The arthritis in my knees does not seem to tolerate "down" nearly as well as "up," and my right big toe presses painfully against my boot. The downward journey, even carrying little gear, just seemed endless.

I eventually made it back to camp at roughly 2:00 p.m. and we celebrated with the normal "potato soup" at around 3:00 p.m. I was exhausted, and the thought of the following day having to pack up all my gear and leave the horses behind for the climb to high camp was almost unbearable. I am almost positive that I told Richard in the past that I would only climb henceforth if I did not have to carry my full pack. At that point, I still had a smidgeon of hope that at least some of my gear would be carried by the Kurds, but it was not enough to allow me to sleep peacefully that night. Saim had to leave to go back to town to help a military expedition climb the other side of the mountain, and the horses had returned to lower pastures as well, so everything that went with us from that point on had to be carried by people—unfortunately I was one of the people. That evening, I did a lot of soul-searching, wondering again why I was on this quest—wondering if I was in the Lord's perfect will, or totally out to lunch. Could it be both? One may not necessarily preclude the other, I suppose.

September 16, day five

The climb to high camp can best be described as painful. Though it is only around fifteen hundred feet in elevation gain, it is seemingly endless piles of nearly vertical heaps of volcanic boulders. The last time I made the climb I swore I would never again do such a thing. Maybe the only way to prevent the mental disease that causes me to forget and return is to throw or give away all my climbing gear. Somehow the agony seems to fade with time—only to return in full fury when I am in process.

In the end, the climb "only" took three hours, but they seemed like days. It would not be such a bad climb if it were not for the extra seventy pounds of backpack and gear. I think my pack was probably the heaviest that I have carried up Ararat because I had the climbing rope, my own tent, cooking fuel, stove, all my own food, all my climbing gear, and everything else I normally carry.

I was actually astonished that I was the first to arrive at the high campsite, but in all fairness, the Kurds carried heavy gear slung in duffel bags over their shoulders—very awkward when climbing steep boulders. Ahmet, the mountaineer, had so much gear in his two backpacks that he would carry one load up a ways, put it down, and go back to get his second pack. He was sixty-one years old and he carried probably eighty-five pounds of gear—making two trips to my one, he was only an hour behind me in getting to camp.

The 2005 Search For Noah's Ark

I had set up my tent and gone back a couple hundred yards to the stream to get water when I spied Ahmet and Richard cresting the farther ridge. I filled up my water bottles and decided to go back to help Richard with his camera pack. He refused help, however, insisting that I go back and carry Ahmet's second pack from where he had left it. Luckily for me, it was the lighter of the two packs he carried, and I "only" saved him a few hundred yards. I hope he appreciated the help because I was already exhausted and it had been a struggle to fight the discouragement and exhaustion the entire way. I tried to sing praise songs and ignore what was ahead (one step at a time ...), but often I could not get enough oxygen in my lungs to keep singing and hiking at the same time! After the last two days of vertical climbing, my legs and knees were especially sore. I remember hoping and praying that we could finish our task within the next three days rather than prolonging it for another ten as Richard seemed to envision. My willpower to go farther was going to have to rest for the evening in order to regain some momentum for the following days' search. Knowing that we were fin-

Author resting in the plush surrounds of "high camp." [Author's photograph, September 2005.]

ished climbing for the day did, however, in and of itself, help to bring a smidgeon of hope.

September 17, day six

The "slope" upward from high camp looks unbelievably steep, but without having to carry the full pack, I remembered the climb from two years prior as not being as impossible as it looked. My goal for this day's climb was to first make it back to the area of the "black ice" that I was unable to access two years earlier. It was there that I spied a large area of "black ice" that I tried to photograph through the digital binoculars—both photos that I attempted turned out blank upon my return to the U.S. (in spite of the fact that every other photo turned out fine). Then, after returning to the site to solve the dilemma of the black ice one way or the other, I planned to work my way to the edge of the canyon and climb along the edge to see what lay within.

Ahmet, the Turkish mountaineer, was supposed to stay with me, and Ahmet Abas was to accompany Richard. I left for the climb at a few minutes past 8:00 a.m. with Ahmet following. Once we got to the top of the first ridge, however, I found that my "partner" had already gone his own route, and he was nowhere to be seen. There were more "valleys" and ridges to cross than I had remembered and the terrain was anything but friendly. It seemed to be considerably steeper and more filled with loose boulders and avalanche debris than I had remembered as well. Of course, if I had adequately remembered all the details of my previous climb, I kept reminding myself, I would not have been there at all! It can be a bit disconcerting to realize that virtually every boulder you step on has arrived at where it is by rolling down from somewhere else—and more than just occasionally I would step on what appeared to be an enormous and very secure boulder only to have it suddenly roll out from under my feet.

I crossed three or four such fields, long since having given up hope of reuniting with Ahmet, before I finally got to the canyon of interest. It now seemed from my vantage point that the area of interest from two years ago was more

The 2005 Search For Noah's Ark

melted out and there were large crevasses with no visible structure showing. From where I stood, I realized that it was actually far more treacherous to climb down to the edge of the canyon than I had hoped or assumed. I started down in a couple of different areas only to be trapped by a drop-off or by crumbling rock that was too dangerous to attempt without gear or a partner. On more than one occasion, I was somewhat concerned about getting back. Actually, "frightened" might be a better word. I finally found one place where, hanging by my ice axe, I could drop a couple of feet and make it down. As I was suspended from my axe, and about ready to make the drop, a little voice inside said, "How will you get back up?" I hung there for a few seconds contemplating my fate and realized it was the better part of valor to listen to that little voice.

I began working my way up the ridge to see if I could get a view of the upper part of the canyon from higher up and I eventually saw Ahmet some ways above me. I worked my way up to him and we continued to climb until we were well above fifteen thousand feet. But once we were near the mouth of the canyon, there were some impossibly steep ice fields to cross and a lot of crumbling rock just waiting for a chance to roll down the hill. Since we had no gear with us, no ropes or harness, no crampons or ice axe, I realized it would be foolish to go on. Ahmet, however, asked to borrow my digital camera—if I had known what he intended, I would have courteously refused!—and he headed for the lip of the canyon anyhow. More than once, I thought he was a goner (along with my camera!). I began to intercede fervently (more for Ahmet than for the camera, but I did include the camera!). When he got to the edge, he took a few shots and then disappeared over the edge for quite some time.

When he had been gone for over ten minutes and did not reply to my calls, I began to wonder what I would do if he did not return. How long should I wait before I left to find Richard and come back with the necessary gear to go and look for him?

Thankfully, he eventually showed up, but he had a far more difficult time returning—it is always easier to climb a difficult area than it is to descend, if only because coming down does not afford you the view that you get on the way up. When he finally traversed the last steep ice field, he stopped and mopped his sweating brow with pieces of snow and ice. I shook his hand when he returned (and quickly took possession of my camera).

It was not until much later that I was able to look at the pictures that he had taken with my camera and I discovered one seemingly large, rectangular object of considerable interest. Had I known at the time how to utilize the "zoom" feature in observing the pictures stored in the camera, I believe I would have made a strong case to Richard for returning to this object with rope and gear for a descent. As it turned out, it was not until returning to America that I was able to view the object with more detail ... and interest (see photo at end of epilogue).

We made it back to camp around 1:00 p.m., so it constituted about five hours of climbing and exploration. I decided to break out my camping stove and get rid of some of my freeze-dried food supplies. It seems I have carried them back and forth from Turkey for many years now. I made some "sweet and sour pork with rice"— a nice change from the daily fare of Kurdish rice and potato soup. At a moment of inspiration I poured the remainder of the boiling water into my water bottle and let it sit while I put things away and straightened up. After eating, I used the hot water in my bottle to shampoo and rinse my hair. It is undoubtedly a good thing that no one had a mirror, but it felt good to have clean hair for the first time in a week!

Richard had been planning to climb again the following day and go above the Ahora Gorge and across the ice in order to come down the east ridge. I had tried to dissuade him because the Kurds had no climbing gear, and their "packs" consisted of either duffel bags tied to their shoulders or burlap bags—an impossible way to climb and cross a glacier. After returning from the day's climb and looking at the seemingly sheer ice fields above the gorge, I came to the conclusion that it would be foolhardy to attempt that route with the help and equipment that we had. I realized that Richard would be unlikely to carry much gear if we attempted that route, and the Kurds were climbing in sneakers and rubber slip-ons while trying to carry heavy bags slung over their shoulders, all without any proper ice gear. I had no desire to rope together with others that had no crampons or ice axes—in short no way to stop a slide should a fall occur. Thankfully, after coming back down, Richard had come to the same realization, and he even declared the climb "impossible" (a word usually not even in his vocabulary!). However, the only apparent alternative was to descend to around eleven thousand feet and then work our way around the mountain to the gorge. And no one even knew if it was possible to do so. We had no idea whether we would encounter impassible ravines or impossible cliffs along the way. But it seemed we would either have to attempt it or return to base camp and give up our time on the mountain. In addition, the change in plans meant that we would now have to carry all of our gear around the mountain for at least three or four days without access to any horses and a limited access to clean water at the lower elevations.

We had had no sign of any promising sites, or areas of potential interest (other than the photos on my camera that remained untapped) and it looked as if we had ruled out several areas altogether. If the Abas family really had some knowledge that this was indeed the "ark area," it would seem that the ark must be now buried completely in avalanche debris. And I further bemoaned the fact that coming down with a full pack would more fully expose the painful arthritis in my knees and the fact that I am not getting any younger. After a week on the mountain, my digestive system had finally surrendered to Turkish cuisine. I had been careful with my water supply and I could only attribute the digestive failure to the apple that I had unthinkingly accepted from Saim two days before. After the first signs of impending doom, I immediately took two Imodium AD tablets, and a few hours later took two more. Then, about 3:30 p.m., I realized that we were now at the fun part of the day—the climb was over, supper eaten, and there was nothing to do for the rest of the night! They say, "It's not over until it's over," but at that moment, I seriously wished it already was!

September 18, day seven

I think that, for the first time on this trip—maybe on Ararat!—I was able to get some sleep during the long night. Before surrendering to sleep, however, I wrote in my journal for some time and then read about King David. I probably caved in and tried to sleep before 8:00 p.m., but I did not really keep track. However, I awoke around 10:30 and had to exit in order to use the "facilities" and I realized that I had been in the midst of a most pleasant dream. I awoke twice more in the night and had been dreaming both times, so I was very grateful for the rest. When I awoke for the last time, it was around 3:00 a.m., and it proved to be impossible to return to sleep mode after that.

I arose around 6:30 a.m. and started organizing and packing for the trek down. We had learned the day before that we were not going to be able to get horses to carry our gear around to the Ahora Gorge. For me, that was a most disheartening revelation—because of arthritis in my knees, going down is much more painful than climbing. I knew it would be one thing to carry my pack down to where the horses could meet up with us, and an entirely other matter to carry my full gear for two or three complete days to get around the mountain and into the gorge. If I had only known the reality of the terrain and what was ahead, I believe, in retrospect, that I would have taken my own gear and made my way slowly back to Dogubayazit. In fact, I am quite sure from their indications to me—in spite of the language barrier—that the Kurdish shepherds would have gone with me!

Every time I believe I have experienced the worst that Ararat has to offer, circumstances prevail to prove me wrong. If you could look up "debilitating agony" in the dictionary, there would have to be a photograph of Ararat attached—it is not for naught that it has been named "Agri Dah" or "painful mountain." My pack was roughly seventy pounds and coming down was agony. For probably the first time in my climbing career, I brought up the rear—and not by just a little bit. "Luckily," just as I caught up with the others finishing their break, I stumbled, lost my balance with the pack, and fell. When the others came to my assistance and tried to help me off with the pack, they realized it was overloaded, and young Ahmet relieved me of the tent, and Mehmet of my sleeping pad. It probably amounted to less than ten pounds, but I was grateful for every ounce.

I eventually lost track of how many rock slide, avalanche fields we crossed. Every time we went down one agonizing, rock-strewn hill, and crossed the impossible terrain, only to have to climb to the ridge on the other side, I prayed that it would be the last. But it never was! For eight excruciating hours it was just more of the same, and I fell farther and farther behind. Many times I lost sight entirely of everyone else and I could only keep track of the general direction they were headed and pray that they would ultimately realize they were missing one of the group. It would be mentally and physically debilitating when I would eventually catch sight of them taking a break some distance ahead only to see them pack up and leave before I ever even caught up. Unfortunately, that meant that I never had time to take my pack off and rest, or I might lose track of them completely. Climbing to high camp, even with my heavy pack, I was first at the top and had even set up my tent before the first of the Kurds arrived, but now coming down was an entirely different story.

In addition to the painful condition of my knees, as strange as it may sound, the worst of the pain was in my right big toe. On two previous climbs, coming down had eventually cost me the toenail of the same big toe. It may seem like a petty issue, but it actually becomes excruciating to walk downhill, and every step becomes tentative just knowing what it is going to cost. The last couple of expeditions, I apparently managed to avoid such a loss because I did not have to carry a full load for as long or as far, and after the horses took over the gear, I was able to wear sneakers instead of my heavy boots. When it eventually became clear halfway through the day that I was going to lose the toenail again, I tried walking sideways and even backward, but it just was not practical for any length of time. Conceding the loss of the toenail was not nearly as bad as enduring the pain with every step.

At one point, crossing an avalanche field of large boulders, the boulder that I stepped on shifted, and I tumbled headfirst down the boulder field. I managed to shout out, "Richard!" just as I pitched, but I recall thinking I was a goner. It even flashed instantaneously through my head that there would be no way for them to transport my body (even if it was not completely dead!) back to civilization. Somehow, in midair, my body reflexively twisted and I landed mostly on my backpack, which bore the brunt of the collision with the rocks. I did manage a nice knot on the skull, a scrape on the wrist, and another knot on my knee, but after lying in a heap for some lengthy seconds trying to assess the damage, I realized that the Lord had been gracious and I had no severe injuries or broken bones. Unfortunately (for the others), that tumble hobbled me even further—and with my aching knees, bruised and failing toenail, bruised and battered knee, I knew I could scarcely afford another serious fall (I actually kept track after that, and though I fell three more times, the rest of the falls were in more mundane conditions and did little damage other than to my already bruised ego!).

Late in the afternoon, I finally espied what I (erroneously!) assumed was our final destination for the day.

The 2005 Search For Noah's Ark

I realized that the others were quite some distance ahead, but things are always farther than they seem, and this time it even seemed a long ways. By this time I was also agonizingly slow! I thought I would never make it and embarrassingly, I actually began to pray that one of the Kurds would come back and offer to carry my pack to camp. It never happened, but that was partially because they had only put down their packs in order to scout ahead and find a place to actually set up camp. It took me well over an hour to work my way down to their location and just as I was beginning to think that I might actually catch up with them before night fell, they all loaded up their packs and headed off again. I think if I had not been so exhausted I might actually have cried, but crying would have taken too much energy (and wasted precious water!).

I managed to hail Richard before he too left, and asked what was going on. He explained that they had found a "better" campsite, closer to water, and "not far off." By that time, even five additional steps seemed like a long ways off, and Richard proved to be wrong in his assessment of "not far off." There ended up being yet another boulder-strewn, precipitous, avalanche canyon to descend in order to get to the valley floor, and even then it was quite some distance to where the tents were being set up. Climbing slowly down, I had to pause once or twice to watch the boulders careering down at me so that I could decide if I needed to "duck and cover" quickly, but nothing came near enough to cause panic.

When I finally reached the somewhat more level plain where the tents were being set up—though there were still a lot of rocks and undulating terrain to traverse—I was just beginning to once again think I might make camp when Mehmet actually came back to carry my pack the final hundred yards to camp. He was wearing his worn-out, rubber, slip-on shoes, and he did not even bother to buckle on the pack, but even with no load to carry, I still could not keep up with him!

The Kurds had a pot of sauceless spaghetti going when I arrived, and as soon as we started to eat, a few chunks of hail began to fall. Looking up at the mountain and seeing the lightning and hearing the thunder made me all the more grateful for God's provision for weather up until this point of our trip. Each day we stayed on the mountain had been sunny and precipitation free. Just as we began the compulsory after-dinner chai (tea), the hail began in earnest. We each dove for our respective tents and then the rains, hail, thunder and lightning came unendingly. It would have been a miserable—not to mention vastly more dangerous—climb down the mountain if the storm had come even a few hours earlier.

Sitting inside my tent and enjoying the rest and safety of shelter also made me realize how difficult—if not impossible—the next day's hike might be. As I began to journal the day's events, I realized that it was not yet even 6:00 p.m., supper was over, the storm was unrelenting, and I was confined to quarters for at least the next twelve hours—the part of the day that I find most galling. I found myself thinking, "If only I could sleep for twelve hours!" My shoulders were quite sore from the weight of the pack, my knees were aching, as well as knotted, my head was throbbing from the lumps it incurred, my wrist was sore from the abuse it received in my fall, and I could only lie there awake and listen to the fury of the storm. Not for the last time it occurred to me that a nice body massage would be well received about then.

September 19, day eight

Looking up at the mountain above us in the morning light, I saw that it was buried in snow, and it occurred to me in even greater clarity how miraculous had been the timing of our descent and the storm. Had we arrived or left even one day later, or the storm come even a few hours earlier, or had we tried, as we had planned, to climb up over the ice to the top of the gorge, we would have been buried in snow and ice. Even had it been possible to extricate ourselves, coming down over rocks

The 2005 Search For Noah's Ark

covered in snow and ice would have proved extremely dangerous, if not impossible. As I painfully crawled from my tent, the thought of facing another full day of the same conditions was almost unbearable. My thighs, knees, shoulders, and especially my right big toe(!) were all quite painful. I decided that I would have to take the risk of wearing my sneakers instead of my hiking boots in order to avoid the excruciating pain of continually stubbing my now loosening toenail against the heavy boot. I also made the decision that at the end of the climb I would give the boots to young Ahmet. Though they were originally $200 boots (for which I paid $125), that had been eight years previously, and they have now cost me three toenails (from the same toe!). In fact, the only time I have ever worn the boots is on Mt. Ararat, and should I ever return it will be with better-fitting boots. Ahmet was originally wearing some beat-up, worn-out old shoes when we started our climb. At high camp, after Saim had asked him to stay on and help "porter" our gear and cook for us, Richard had taken pity on Ahmet and given him a pair of his own sneakers. I was surprised that a size 9 fit him because he is quite tall. However, by the time we reached camp on the eighth day, even Richard's sneakers were falling apart.

There is probably no effective way to capture the eighth day in words, and maybe even less reason to do it, because it would make for almost as painfully slow reading—or at least repetitious—as was the excursion. It began with discussion about what route to take. Turkish guide Ahmet wanted to climb up (for, it seemed, another photo op for his business brochures) to the top of the gorge and then back down. But without climbing gear and ice equipment, the Kurds would have been more of a liability than an asset. In addition, camping would have been nearly impossible and we could not have made it up and down in a day. Sanity eventually (or so I thought) prevailed and we agreed to continue circumventing the mountain until we reached the Ahora Gorge Valley where we could camp and meet Saim coming up with the horses and new provisions.

Dinosaurs on the Ark

Nothing on Ararat is ever what it seems. From far above, what appears to be a gently sloping, grassy valley, is actually rock-strewn and full of ravines and impassible crevasses when up close.

We headed up the nearest ridge only to find another valley and ridge on the opposite side—and a very steep and treacherous descent into the valley below. Another "discussion" ensued between Richard and Ahmet (Ahmet actually knew very few English words, but he seemed to know "photo" quite well and it appeared that he wanted to climb to the top of the ridge in order to have Richard photograph him there). Richard eventually realized that it would be a futile waste of both time and our very limited energy reserves, and, instead, we headed down into the next valley. The descent proved quite challenging as the snow from the night before was still clinging to the shadows on that side of the ridge, and footholds were precarious and extremely narrow. I tried hard to plant my sneakers very carefully and firmly, and engage my ski poles securely, not to look too far ahead, and to pray fervently. One or all of those prevailed, and I survived without a fall, but it was, nevertheless, a harrowing experience.

Arriving at the valley "floor" proved to be of little relief because it was filled with rock-strewn gullies that had to be crossed before we could begin climbing to the next ridge—which appeared even higher. At least we had the comfort of knowing that we would then be able to begin the descent into the Ahora Gorge. Or so we thought at the time. When we finally did crest the ridge on the other side, what faced us appeared no different than what we had just traversed, and there was no sign of the gorge! I eventually lost count of how many ridges we crested only to find another valley and another ridge on the other side, but it began to seem never-ending. And each ridge that we had to descend caused my aching knees to beg for mercy and my throbbing toe to force me to descend sideways. After somewhat more than four hours of such up-and-down agony, we finally did come to the edge of the gorge—and there was no way down! It was probably

The 2005 Search For Noah's Ark

at least a thousand feet to the valley floor and impassible. The only option was to follow the course of the gorge and head down toward the valley and Ahora Village.

Again, from above, the slopes looked gradual and even pleasant, but they proved to be anything but. I thought that my legs would surely be in pain a week after finishing the climb! For me, the descent was now excruciating, and it became so painfully obvious that I could not keep up, that a couple of times, one Ahmet or the other came back to carry my pack for a short distance. Under most circumstances, it would have been a matter of pride that I insist on carrying my own gear, but there were times that I saw how far ahead the Kurds were, and I saw the obstacles in front of me, and I began to despair of ever reaching the bottom.

In the end, they only actually relieved me of my load about three or four times, and they always waited until I was within the final hundred yards of where they sat—and usually when the climbing conditions were no longer difficult.

After a couple of hours of steep and seemingly endless descent, we came to another impassible cliff directly in our path, and it began to occur to me that we did not actually know if it was even possible to reach the bottom by this route. However, that possibility led to the ramification that, if we could not find a path down, the only option would be to retrace our path of the last two days! As soon as that idea came into my mind, I did my best to repress it and keep my thoughts on taking the next step only. At this particular juncture, we came to an impassible precipice and we ended up having to retrace our path somewhat and climb to the adjacent ridge and then try to work our way down the opposite side.

Finally, far ahead, on what appeared to be the grassy knoll of a shepherd's camp, Richard and I saw the Kurds stop. Even though, from our vantage, it seemed impossibly far away and agonizingly "down," I began to feel a smidgeon of hope arise that I could eventually get there. However, long before we actually got close, we saw them pick up and head down the side of the hill. Initially, we assumed they were headed to obtain water for our camp, but they never returned—and we, luckily, had no idea how far we would yet have to go before we found the actual camp! Turkish Ahmet had awaited us at the knoll, and he even came back to carry my pack (when we were within fifty yards of where his pack sat!). This time I refused because by that time we were on fairly level terrain, and I (wrongly!) assumed we were finished for the day! Had I known we still had hours to go...

After a few minutes rest, Ahmet somehow made us understand that we had to descend yet another hill, and cross another ridge before we would find camp. When Richard and I finally crested what we assumed would be our last ridge of the day (long since having lost sight of Ahmet), we were convinced that camp would be in sight. It was not. In fact, we saw nothing of the Kurds, our Turkish mountaineer, or of camp.

When we finally eventually spotted the Kurds waiting in the distance, they appeared to pass up several nearby ideal camping spots—near running water, no less!—and headed off again when Ahmet reached them. The discouragement was definitely palpable! I think Richard was angrier than he was discouraged, because though he was financier and "boss," no one had bothered to consult with him for directions or advice.

On the other hand, I was mortified. Long before that point, I had felt on the brink of succumbing to total exhaustion. I had to force myself to not look ahead, nor try to visualize the camp, nor even hope for it, but only to look for a spot for my next step. And I prayed that I could keep going.

Once again, there were a lot of gullies, rocks, and ups and downs along the way and it seemed to be an eternity, though we finally reached camp a little after five o'clock in the evening. It had been eight and a half hours of constant climbing with no food along the way—other than a couple of dry cookies—and very little water. At higher

The 2005 Search For Noah's Ark

elevations, with the streams coming down from the ice-melt, I have never had a problem drinking the water unfiltered or untreated. Now, however, camp was set up in the midst of a sheep pasture, and the available water ran through a small irrigation ditch. Even though I had been out of water for several hours, and I was dying of thirst, I promptly sat down and boiled three quarts of water for the next day. The Kurds thought I was a bit daft because they were quite happy with the plentiful "su" and could not conceive of a reason for taking nice, cold water and making it hot, only to store it. On the other hand, I have contracted giardia before and it was quite unpleasant to say the least. I placed one of my newly created water bottles in the stream, hoping to have cold water by morning, and then had a bit to eat.

After the evening's repast, the two Ahmets decided to walk (as if we had not already walked enough that day!) to the nearby village of Ahora to get some "treats" (it was probably two miles away). I managed to "order" two orange Fantas, and I contributed twenty lira from my Bible (about $16). I also gave the younger Ahmet my boots to wear—he grinned, put them on, and muttered, "choke guzell," which, roughly translated, means "Ah, very good!" I also loaned him my headlamp for the dark return. When they finally did return after a couple of hours, they had brought flatbread, yogurt, eggs for breakfast, a few other Kurdish delicacies, and soft drinks. I managed to inhale an orange Fanta and a Coke, and then to secret two extra Fantas in the stream to preserve them, as well as to make them cold for breakfast. As I went to my tent, Ahmet came to return my boots and he could not contain his grin when I made him to understand that they were a gift for him to keep (no more lost toenails for me!). I gave the headlamp to Mehmet, but I waited until morning because I had reading and writing to do in the tent that evening. Eventually, I decided to also leave my crampons with Saim for his "expedition" business.

That night I made the decision that it was time for me to go back to Dogubayazit, recover for a day or two, and then return to the States. Richard was determined to stay on and then, when re-outfitted by Saim, head back up into the gorge.

I actually stayed awake that night until maybe 10:30 p.m. (I have been awake far past that hour on Ararat, but never when I was trying), and I believe I managed to sleep sporadically throughout the night. It was, however, hard to find a position where my thighs, knees, and big toe did not hurt.

September 20, day nine

In the morning we got up leisurely—it was nice to have relatively warm conditions finally and I could go outside in a T-shirt (we had come down to around sixty-seven hundred feet). For breakfast, the Kurds made some kind of "egg soup" and began dipping into it with flatbread and fingers, and they polished off the yogurt, cheese, and leftovers from the night before as well. They insisted that Saim was coming at 11:00 a.m. with the Lada to pack up those who were going back to town. Richard had seemed very disappointed—and even angry—two days before when I told him that I did not want to extend my stay in Turkey beyond the twenty-fourth (which meant that I would have to be back in Dogubayazit by the twenty-second at the latest). However, the last two days of "climbing" almost demoralized even Richard the indefatigable! By day nine, he seemed resigned to the fact that I had other obligations that needed my attention in America.

After breakfast, and after packing up my gear, it seemed we would have a two- or three-hour wait for Saim to arrive, though, on "Turkish time," it could always end up being much longer. I spent my time reading from my Bible and praying, but time moves very slowly when you are sitting and waiting ... and waiting (ever been in a doctor's waiting room for an hour or so?). I was grateful that my knees and thighs were not as painful as I had anticipated—though it was not easy to rise! As I took down my tent, I noticed that the small irrigation ditch that ran by the front of my tent was now empty,

The 2005 Search For Noah's Ark

and I surmised that the shepherds were able to direct the water to various areas at different times. As I looked at the mud, I noticed that barely three feet from the front of my tent a large cat had recently walked through the mud of the ditch. I assumed it was a cougar because I had read stories of such creatures existing on Ararat (later I learned from a friend's research that it could have been a Caspian tiger—though most think they are now extinct in the area—or a Persian leopard). At first Richard was dubious of my claim, but as I was pointing out the tracks, one of the shepherd boys also looked and with eyes wide, made his hands into claws, and made hissing noises to indicate a large cat. I was grateful that I had not encountered it a few hours before when, shortly before sunrise, I left the tent at night for a bathroom break. I am not sure how I would have reacted to a large "cougar" cruising for breakfast ... or he to me. But I am quite sure I would have been in no condition to tangle!

Looking back at the Ahora Gorge from our final campsite. A photograph cannot seem to capture the vastness of the "Painful Mountain." [Author's photograph, September 2005.]

After we'd sat around the remnants of our campsite chatting for a while, several Kurdish shepherd children showed up to entertain us—or to be entertained, though, more likely they were looking for handouts. At a whim, Richard began asking them if they knew of "Noah's Ark" (though he pronounced it in Kurdish). Several pointed to an area of the mountain and said, "Yes." Initially, it seemed to me that several pointed to different areas (there were about five children), but when Richard pulled out some photos, they all seemed to agree on one area—the very region that Richard had wanted to go to based on another "eyewitness" account. Admittedly, in one of the photographs, the area was circled, and in another, there was a red arrow pointing to the spot, though it seemed that at least some had pointed to the same area on the mountain before seeing the pictures.

I must admit to a degree of astonishment, even now looking back, at the transformation that began to take place in my mind. The night before, I had been exhausted and in pain, and thoroughly convinced that I had seen the last of Ararat. All of a sudden, the thought that I was less than a two-day climb (and with horses to carry the gear to our campsite!) from at least a part of the ark caused me to begin to agonize over what I should now do. The pain and exhaustion of the night before were already greatly diminished and I reasoned that by resting the remainder of the day, I could survive another "hike" up the gorge. I had given away my mountaineering boots, but I began to reason that once Saim arrived to translate, I could borrow them back from Ahmet for a few days. In Richard's favor, he did not pressure me or try to dissuade me from leaving, but I knew that he wanted me to be with him for the "big event." I walked around and tried to find a quiet place just to pray. It was clear that Mehmet and Ahmet were not willing to stay on—they were both quite anxious to head back to Dogubayazit and to have me go with them. I think they may have felt that if I stayed, they might have been unable to leave as well. I started going through all the reasoning, and I kept wondering what it would be like to find out in a week or two that I had been so close and had given up. In the end, I tried to pray and just allow God's peace to speak to my heart. All that came to me was, "Go home." I tried arguing with that "thought" and testing it to see if it was just exhaustion, or cowardice—or both!—but in the end I achieved the most peace of mind just by acknowledging that inner voice. I prayed that Saim's arrival (if and when!) would confirm and solidify that message.

In the end, Saim did not finally arrive until after 2:30 p.m. (and that *is* roughly 11:00 a.m. in Turkish-speak!) and he was dressed in street clothes—certainly not headed for the mountains! There were no horses carrying gear and no supplies, and it was clear that he had come only to take us back to town. He did speak with the children to find out what they actually knew, but according to his version they actually knew nothing and were only hoping for a payoff. Ahmet, our Turkish mountain guide, had packed up his own gear, hired some shepherds with donkeys, and headed off on his own around 11:00 a.m. He did not let us know where he was going, nor did he say his good-byes. We then found that, apparently, when Saim had reached him on the cell phone two days previously, Ahmet had told him that I had fallen and was too badly injured to continue, and that he should just come and get us. I had wondered at the time why Saim had not talked to either Richard or me on the phone. But we found out that he had insisted that he speak with one of us, and Ahmet had told him we were "not around." Richard had continued to insist that he get to talk with Saim, but Ahmet continued to feign ignorance and just hung up before either of us could talk. Saim, being very worried about me, had not outfitted another expedition, but had come to retrieve us instead. In fact, Richard was initially insistent on staying with the camp until Saim returned, but Saim would not hear of it. By himself, Richard would not be "safe" so near the village and his things could be stolen as well. In addition, Richard had no drinking water, nor did he have sufficient supplies to remain on his own for a couple of days. So we bundled up the last remaining tent and all the gear and headed down to Saim's Range Rover.

Passing through the outskirts of Ahora Village, we even somehow managed to reconnect with Ahmet, so we ended up with four of us in the backseat along with Ahmet's gear and backpack—the makings of a very uncomfortable two-hour ride back to Dogubayazit (though I kept forcibly reminding myself, "Be thankful—this is far better than the conditions of the last two days!"). When we finally arrived back at the campground in Dogubayazit, we feasted on chicken and lamb—and quarts of orange Fanta!

The following morning, I was amazed at the recovery in both body and mind, so I walked five miles into town in order to check e-mails ($0.40 for half an hour!) and get a few essentials. The next day, I left by minibus for Kars and a flight to Istanbul, where I had a day to kill before beginning the long trek back to the U.S.

It has often seemed that each trip to Turkey has produced more questions than answers, along with a number of what-ifs: What if Paul had not fallen ill in 2001? What if Paul and I had had adequate water and good weather in 1999? What if Mustufa Ozturk had been telling the truth about what he had seen and he had made good on his promise to accompany me the next year? What if I had been able to find the Armenian farmer who owned the book and map to the ark in 2005? What if Richard and I had been able to follow the Lake Kop route in 2000? What if we had been able to convince the Abas shepherds to tell us all they really knew in 2003? What if I had been able to utilize the zoom function on my camera and get a good look at the rectangular protrusion before leaving the area in 2005? What if, what if, what if? And yet, as I look back, there has been miraculous provision, protection in an often hostile environment, and, I believe, some answers as well. Answers to the remaining questions regarding Ed Davis's account. Answers regarding the position of the military in Turkey. Answers regarding the mountain itself. In fact, after returning to America and being able to examine the photographs from the 2005 expedition, I began to wonder if maybe we had not been given the ultimate answer to our entire quest—the location of at least a piece of the ark? Rocks are jagged and irregular, but there is one very large, ice-encrusted, rectangular protrusion jutting prominently from the ice in some of my photos that requires further examination.

EPILOGUE

The weather is changing. Some cry "global warming!" Others debunk such claims as political fodder, but a pastor friend of mine told me in 2005, "The Lord is changing global weather patterns to reveal the thing you seek." Whether it amounts to global changes or just seasonal, regional changes, it seems that glaciers in certain areas are indeed in recession. It may not be long before that which is hidden on Mt. Ararat can no longer remain so. In the meantime, the search will go on. I believe that I need to make one more trip to Armenia to try and verify once and for all whether or not there is a book with a picture of the ark and a map to its location. If such a thing exists, it would be a huge boon to the search for Noah's Ark. Either way, there is an area on Mt. Ararat with a curious rectangular outline in the snow that certainly deserves a closer look.

Most rocks are jagged and irregular, but there appears to be something symmetrical and rectangular jutting from the ice in this photo. [Author's photograph, September 2005.]

We'd love to have you download our
catalog of titles we publish at:

www.TEACHServices.com

or write or email us your thoughts,
reactions, or criticism about this
or any other book we publish at:

TEACH Services, Inc.
254 Donovan Road
Brushton, New York 12916

info@TEACHServices.com

or you may call us at:
518/358-3494

LaVergne, TN USA
22 February 2010
173848LV00001B/2/P